Measurement and Geometry

Contents

Introduction

Building a solid foundation in math is a student's key to success in school and in the future. This book will help students to develop the basic math skills that they will use every day. As students build on math skills that they already know and learn new math skills, they will see how much math connects to real life.

This book will help students to:

- develop math competence;
- acquire basic math skills and concepts;
- learn problem-solving strategies;
- apply these skills and strategies to everyday life;
- gain confidence in their own ability to succeed at learning.

Students who have self-confidence in their math skills often do better in other school areas, too. Mastering math helps students to become better learners and better students.

Ensure Student Success in Math

This book contains several features that help teachers to build the self-confidence of math students. This book enables the teacher to:

- reach students by providing a unique approach to math content;
- help students build basic foundational math skills;
- diagnose specific math intervention needs;
- provide individualized, differentiated instruction.

Assessment. An Assessment is included to serve as a diagnostic tool. The Assessment contains most of the math concepts presented in this book. An Assessment Evaluation Chart helps to pinpoint each student's strengths and weaknesses. Then, instruction can be focused on the math content each student needs. Each item in the Assessment is linked to a lesson in the book where students can hone their math skills.

Correlation to Standards. A Correlation to NCTM Standards is provided to allow teachers to tailor their teaching to standardized tests. This chart shows teachers at a glance which lessons cover the basic skills students are expected to master.

Lesson Format. Each lesson in the book is constructed to help students to master the specific concept covered in the lesson. A short introduction explains the concept. Then, a step-by-step process is used to work an example problem. Students are then given a short problem to work on their own. Finally, a page of practice problems that reinforce the concept is provided.

Glossary. Math has a language of its own, so a Glossary of math terms is included at the back of the book. Students can look up terms that confuse them, and they are directed to a specific page on which the term is explained or implemented.

Answer Key. A complete Answer Key is provided at the end of the book. The Answer Key includes the answers for the practice problems as well as explanations on how many of the answers are reached. These explanations can be useful to the teacher to explain why students might have answered incorrectly.

Graphic Organizers. Graphic organizers often help students to solve problems more easily. For that reason, a series of charts, detailed step-by-step processes, and various kinds of graphs and diagrams are supplied at the back of the book. Students can use these tools to help them solve the problems in the book or create their own problems.

Working Together to Help Students Achieve

No student wants to do poorly. There are many reasons students may be having problems with math. This book presents a well-organized, straightforward approach to helping students overcome the obstacles that may hold them back. This book and your instruction can help students to regain their footing and continue their climb to math achievement.

Correlation to NCTM Standards

Content Strands

Lesson

Measurement

• understand both metric and customary systems of measurement	1, 2, 3, 4, 12, 13, 14, 15, 16, 17
• understand relationships among units and convert from one unit to another within the same system	1, 2, 5, 12, 13, 14, 15, 16, 17
• understand, select, and use units of appropriate size and type to measure angles, perimeter, area, surface area, and volume	8, 9, 10
• select and apply techniques and tools to accurately find length, area, volume, and angle measures to appropriate levels of precision	9, 10, 26, 27, 38, 39, 40
• develop and use formulas to determine the circumference of circles and the area of triangles, parallelograms, trapezoids, and circles and develop strategies to find the area of more complex shapes	9, 10, 25, 26, 27
• develop strategies to determine the surface area and volume of selected prisms, pyramids, and cylinders	35, 38, 39, 40
• solve problems involving scale factors, using ratio and proportion	6

Geometry

• precisely describe, classify, and understand relationships among types of two- and three-dimensional objects using their defining properties	18, 19, 22, 23, 24, 35, 36
• understand relationships among the angles, side lengths, perimeters, areas, and volumes of similar objects	19, 20, 21
• create and critique inductive and deductive arguments concerning geometric ideas and relationships, such as congruence, similarity, and the Pythagorean relationship	28, 29, 30
• use coordinate geometry to represent and examine the properties of geometric shapes	31, 32, 33, 34
• describe sizes, positions, and orientations of shapes under informal transformations such as flips, turns, slides, and scaling	31, 32, 33, 34
• examine the congruence, similarity, and line or rotational symmetry of objects using transformations	28, 29
• use two-dimensional representations of three-dimensional objects to visualize and solve problems such as those involving surface area and volume	36, 37, 41
• recognize and apply geometric ideas and relationships in areas outside the mathematics classroom, such as art, science, and everyday life	18

Name _____ Date _____

Assessment

Solve.

1. 72 in. = _____ yd
2. 2,670 m = _____ km
3. 508 cm = _____ in.

4. 5 km = _____ mi
5. 60 in. + 2 yd = _____ ft
6. 1.8 km − 258 m = _____ m

7. $\frac{1}{4}$ mi = _____ ft
8. 1,672 cm = _____ m
9. 120 in. + 2 ft = _____ ft

10. 3.2 km − 145 m
    = _____ m
11. 5 in. = _____ cm
12. 1.6 km = _____ mi

13. What is the area of a rectangle that is 16 meters wide and 4 meters long?

14. A model racecar is 10 inches long and the scale is 1-25. How long is the real car?

15. What is the area of a rectangle that is 12 feet wide and 2 feet long?

16. A model sailboat is 6 inches long. If the scale is 1 inch = 2.5 feet, how long is the actual boat that the model represents?

Use the distance formula ($d = r \times t$) to find the distance, rate, or time.

17. Sarah was driving for 5 hours. She drove 305 miles. What was Sarah's average speed?

18. How far would Elana travel if she drove her car at an average speed of 57 mph for $3\frac{1}{2}$ hours?

19. In a recent Daytona 500 car race, the winning car had a speed of 133.87 miles per hour. How long did it take the winner to drive the 500 miles?

20. The Iditarod is a dog sled race from Anchorage, Alaska, to Nome, Alaska. One year, a racer traveled at a rate of 105 miles per day for 11 days. How far did the racer travel?

Name _____ Date _____

Find the perimeter and area of each figure.

21.
```
      4 m
   ┌──────
3 m │    ╱
   │   ╱ 5 m
   │  ╱
```

22.
```
   3 ft
  ┌────┐
  │    │ 3 ft
  └────┘
```

23.
```
  ┌────┐
  │    │ 2.2 yd
  └────┘
   3 yd
```

24.
```
        19.72 km
10 km ╲
      │ ╲
      └───╲
       17 km
```

Solve each capacity problem.

25. 10 pt = _____ qt

26. 200 mL = _____ L

27. 2 gal = _____ L

28. 7 pt = _____ qt

29. 2000 mL = _____ L

30. 60 mL = _____ fl oz

Solve each mass problem.

31. 12 lb = _____ oz

32. 10 kg = _____ g

33. 7.5 kg = _____ lb

34. 6 tons = _____ lb

35. 1050 g = _____ kg

36. 100 g = _____ oz

Name _____ Date _____

Identify the lines A and B as parallel, perpendicular, or intersecting but not perpendicular.

37. 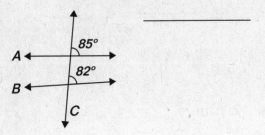 _____

38.  _____

39.  _____

40.  _____

Identify each angle as acute, right, obtuse, or straight.

41. 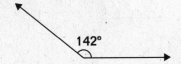 _____

42. _____

43.  _____

44. 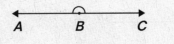 _____

Find the supplement for each angle.

45. 45° _____

46. 72° _____

Use this diagram for problems 47 and 48.

47. Identify all drawn radii of circle O.

48. $\overline{MO}$ is 4 cm. Find the circumference of circle O.

Find the area of each figure.

49. a circle with a diameter of 4 yd

50.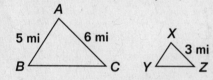

Use the similar figures in this diagram to answer questions 51 and 52.

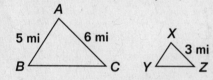

51. How long is side XY?

52. What angle corresponds to ∠Z?

Find the coordinates of each point after the given transformation.

53. Where is the image of point (2, 4) after a (−3, 4) slide?

54. Where is the image of point (5, 3) after a flip over the x-axis?

55. Where is the image of point (2, −1) after a 90° clockwise rotation around (0, 0)?

56. Where is the image of point (−1, −4) after a dilation with its center at (0, 0) and a scale factor of 2?

Name _____ Date _____

Find the volume of each figure.

57.

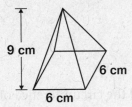

9 cm

6 cm

6 cm

58.

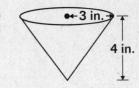

3 in.

4 in.

Find the surface area.

59.

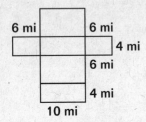

6 mi   6 mi

4 mi

6 mi

4 mi

10 mi

60.

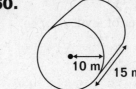

10 m   15 m

61. What is the volume of the rectangular solid?
Use the formula $V = l \times w \times h$.

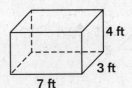

4 ft

3 ft

7 ft

62. Which figure shows the right-side view of this object?

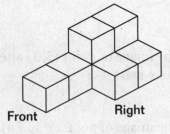

Front   Right

A.

B.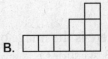

C.

D.

Assessment Evaluation Chart

Note the numbers of the assessment items you missed, if any. Then use the Lesson Review list to find more practice problems.

ITEM NUMBERS	SKILL	LESSONS FOR REVIEW
1–16	Converting Units of Length	Lessons 1–6
17–20	Using the Distance Formula	Lesson 7
21–24	Finding Perimeter and Area	Lessons 8–11
25–30	Customary and Metric Units of Capacity	Lessons 12–14
31–36	Converting Units of Weight and Mass	Lessons 15–17
37–40	Parallel and Perpendicular Lines	Lesson 19
41–44	Angles	Lesson 20
45, 46	Pairs of Angles	Lesson 21
47, 48	Properties of a Circle	Lesson 24
49	Area of a Circle	Lesson 27
50	Area of a Triangle	Lesson 26
51, 52	Similar and Congruent Figures	Lesson 28
53–56	Translations	Lessons 31–34
57	Volume of a Pyramid	Lesson 35
58	Volume of a Cone	Lesson 39
59, 60	Surface Area	Lessons 36, 37
61	Solving Two-Step Volume Problems	Lesson 40
62	Two-Dimensional Views of Three-Dimensional Objects	Lesson 41

Concentrate on any skill section in which you missed one or more problems.

Name _____ Date _____

LESSON 1 Customary Units of Length

You are already familiar with customary units of length. You would measure a pencil in inches. A sheet of paper is about a foot long. A soccer field would be measured in yards, and distance between cities is measured in miles.

Sometimes you may have to change, or **convert,** between units of measure. A roll of tape may be measured in feet but if you need to have inches of tape, you would have to change feet to inches to make sure you had the right length.

When you convert units, you use **conversion factors,** the number that you multiply or divide by.

When you have to change from one unit to another, remember

- When you change from a smaller unit to a larger one, you divide.
- When you change from a larger unit to a smaller one, you multiply.

Customary Units Conversion Factors	
1 foot (ft) = 12 inches (in.)	smaller
1 yard (yd) = 36 in.	
1 yd = 3 ft	
1 mile (mi) = 5,280 ft	larger

Example

One side of a tennis court measures 12 yards. How many feet long is one side?

STEP 1 Identify the units of measure.

yards   feet

STEP 2 Identify the smaller and larger units.

yards ⟶ feet

larger ⟶ smaller

STEP 3 Divide or multiply.
Since you are changing yards (larger) to feet (smaller), you multiply.

12 yd × 3 ft = 36 ft

One side of the tennis court is 36 feet long.

(ON YOUR OWN)

The track team ran 7,920 feet. How many miles did they run?

Measurement and Geometry, SV 0437-9

Practice

Building Skills

Convert the units.

1. 150 in. = _____ ft

2. 2.5 mi = _____ yd

3. 100 yd = _____ ft

4. 2,640 ft = _____ mi

5. 880 yd = _____ mi

6. 5 mi = _____ in.

7. 600 in. = _____ yd

8. 25 ft = _____ in.

9. 3.9 mi = _____ ft

10. 3.2 yd = _____ in.

Problem Solving

Solve.

11. You jump 264 inches in the long jump. How many feet do you jump?

12. At 8.92 feet tall, Robert Wadlow is considered to be the tallest person in history. How tall is he in inches?

13. The largest fish in the world is the whale shark. It can grow to a length of 46 feet. How many yards long is the whale shark?

14. The distance around the Daytona International Speedway is 2.5 miles. What is this distance in yards?

15. The teeth of a dinosaur can be as long as 0.75 feet. What is this length in inches?

16. A tractor trailer is 168 inches tall. It is traveling on a road that has an overpass 11 ft tall. Will the truck fit under the overpass?

11

LESSON ② Metric Units of Length

You may not be as familiar with the metric units of length as you are with customary. However, you probably have heard of a 100-meter dash or a 5K (kilometer) run.

Using the metric system, you would measure a paperclip in millimeters. A sheet of paper is about 30 centimeters. You would measure a soccer field in meters. You would measure the distance between cities in kilometers.

As with customary units of length, you may have to change from one metric unit to another. You use conversion factors to change units.

When you have to change from one unit to another, remember

- When you change from a smaller unit to a larger one, you divide.
- When you change from a larger unit to a smaller one, you multiply.

Metric Units Conversion Factors	
1000 millimeters (mm)	= 1 meter (m)
1 cm	= 10 mm
1 m	= 100 centimeters (cm)
1 kilometer (km)	= 1000 m

smaller ↓ larger

Example

A square cake pan measures 20 cm on one side. How many millimeters is one side of the cake pan?

STEP 1 Identify the units of measure.

centimeters   millimeters

STEP 2 Identify the smaller and larger units.

centimeters ⟶ millimeters

larger ⟶ smaller

STEP 3 Divide or multiply.
Since you are changing centimeters (larger) to millimeters (smaller), you multiply.

20 cm × 10 mm = 200 mm

The cake pan measures 200 millimeters.

ON YOUR OWN

A lake in New York is 2.7 kilometers deep. How many meters deep is the lake?

Practice

Building Skills

Solve.

1. 22 cm = _____ m

2. 2.5 km = _____ m

3. 100 mm = _____ m

4. 5.2 m = _____ cm

5. 0.4 km = _____ m

6. 7,290 mm = _____ cm

7. 0.6 m = _____ mm

8. 347.5 m = _____ km

9. 179,000,000 mm = _____ km

10. 7.6 km = _____ cm

Problem Solving

Solve.

11. One lap in an official Olympic-size swimming pool is 50 meters. How far is this distance in kilometers?

12. The Golden Gate Bridge is 1,280 meters long. What is the length of this bridge in kilometers?

13. A snowboard has a deck length of 116 cm. What is its length in meters?

14. A female pole-vaulter set a world record with a 4,876.9-millimeter high vault. How high was her vault in meters?

15. A runway at Pearson International Airport in Toronto, Canada, is 3.389 kilometers long. What is its length in meters?

16. The wingspan of a male bald eagle ranges from 1.82 meters to 2.16 meters. What is this range in centimeters?

LESSON ③ Adding and Subtracting Units of Length

You are running a track that is 1.5 kilometers long. You are at the 900-meter marker. How much farther do you have to go to finish 1.5 km? What is the total distance if you go an extra 500 m?

When you are adding or subtracting units of length, you first change all of the lengths to the same unit. When all of the units are the same, you can add or subtract the distances.

You can avoid having fractions and decimals in your answer if you convert all of the units to the smallest unit.

4th of July ROAD RACE

You are at: 900 METERS

Example

10 meters + 0.1 kilometers = _____

STEP 1 Change all the lengths to the same unit.

0.1 km × 1,000 m = 100 m

STEP 2 Add.

10 m + 100 m = 110 m

Example

1.3 km − 200.5 m = _____

STEP 1 Change all the lengths to the same unit.

1.3 km × 1,000 m = 1,300 m

STEP 2 Subtract.

1,300 m − 200.5 m = 1,099.5 m

ON YOUR OWN

2.6 km − 520 m = _____

Name _____ Date _____

Practice

Convert all of the units to the smallest unit.

Building Skills

Add or subtract.

1. 45.7 cm − 320 mm =

2. 456 ft − 37 yd =

3. 1,370 cm + 42.7 m =

4. 1.3 mi + 275 yd =

5. 0.8 mi + 4,130 ft =

6. 4,367 mm + 72 cm − 2.5 m =

7. 432 in. + 37.25 ft =

8. 0.6 km − 623 m + 3,200 cm =

9. 16 yd − 31.5 ft =

10. 3.65 km + 537 m − 2.2 km =

Problem Solving

Add or subtract.

11. You walk 30 yards, stop, and then walk 15 feet farther. How far do you walk?

12. Starting 15 kilometers from your house, you drive 3,627 meters toward your house. How far are you from home?

13. What is the length of a 100-yard football field plus two 30-foot end zones?

14. You have two pieces of rope. One is 3 yards long, and the other is 14.5 feet long. What is the total length of rope?

15. Your school is 2.4 miles round-trip from your house. You bike to school and most of the way back, but you stop 500 feet from home. How far do you ride?

16. Three of your friends walk for a charity event. They walk 6,160 yards, 6 miles, and 11,616 feet. What is the total distance your three friends walk?

LESSON ④ Multiplying and Dividing Units of Length

You make the trip to and from school all week. If you know the distance from your home to the school, you can find out the total distance you travel each week. Just multiply the distance to school by 10 (2 trips a day for 5 days). If you know the total distance you travel each week but want to find the distance from your home to school, you can divide the total distance by 10.

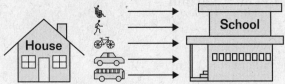

To multiply or divide units of length, first make all the units the same, and then multiply or divide. If you multiply two lengths, you get square units.

Example

$$3.5 \text{ m} \times 225 \text{ cm} = \underline{\hspace{4cm}}$$

STEP 1 Change all lengths to the same unit.

$$3.5 \text{ m} = 350 \text{ cm}$$

STEP 2 Multiply.

$$350 \text{ cm} \times 225 \text{ cm} = 78{,}750 \text{ cm}^2$$

Example

$$5.2 \text{ km} \div 200 \text{ m} = \underline{\hspace{4cm}}$$

STEP 1 Change all the lengths to the same unit.

$$5.2 \times 1{,}000 = 5200 \text{ m}$$

STEP 2 Divide.

$$5200 \text{ m} \div 200 \text{ m} = 26 \text{ m}$$

ON YOUR OWN

$$26 \text{ m} \times 3 \text{ cm} = \underline{\hspace{4cm}}$$

Practice

Building Skills

Multiply or divide.

1. $3{,}760 \text{ mm} \div 40 =$

2. $21 \text{ ft} \times 12 \text{ yd} =$

3. $350 \text{ cm} \times 5.6 \text{ m} =$

4. $0.625 \text{ mi} \times 30 \text{ ft} =$

5. $0.5 \text{ mi} \times 28 \text{ ft} =$

6. $45 \text{ km} \div 600 \text{ m} =$

7. $1.5 \text{ mi} \div 200 \text{ ft} =$

8. $0.2 \text{ m} \div 5 \text{ cm} =$

9. $963 \text{ yd} \div 120 \text{ in.} =$

10. $584 \text{ mm} \times 2.7 \text{ cm} =$

Problem Solving

Multiply or divide.

11. Your classroom is 13 feet long and 20 feet wide. What is the length of the room times the width?

12. You walk in a 5000-meter foot race. If the foot race is divided into 250-meter parts, how many parts do you walk in all?

13. If you jog 4.5 kilometers a day, how far do you jog in 5 days?

14. You are watching a 600-meter relay race. How far is each of the 4 equal parts of the race?

15. A ladder is 14 feet long and is divided into 21 sections. How many inches long is each section?

16. How many 60-foot-long plots can be set up for volleyball courts on a $\frac{1}{2}$-mile of beach?

LESSON **5** Converting Between Customary and Metric Units of Length

You have probably already noticed food labels with both customary and metric units on them. If you measured the edge of a 14-in. book with a metric ruler, you would see that it was 35.56 cm long.

Metric Ruler

To change units of length from customary to metric or from metric to customary, you set up a conversion factor.

Conversion factor is the number that you multiply or divide by to change to another unit of measure.

By learning the conversion factors, you can change metric units to customary ones or customary to metric.

Customary to Metric Conversion Factors
1 inch = 2.54 centimeters
1 foot = 0.305 meters
1 mile = 1.6 kilometers

Example

The distance between two cities is 72 miles. What is the distance in km?

STEP 1 Set up the problem.

72 mi = _____ kilometers

STEP 2 Find the conversion factor.

1 mi = 1.6 km

STEP 3 Multiply.

72 × 1.6 = 115.2 km

The distance between the two cities is 115.2 kilometers.

ON YOUR OWN

The track at school measures 500 meters. What is the length in feet?

Metric to Customary Conversion Factors
1 centimeter = 0.394 inches
1 meter = 3.28 feet
1 kilometer = 0.6 miles

Practice

Building Skills

Convert each unit.

1. 4.5 mi = _____ km

2. 43 ft = _____ m

3. 50.8 cm = _____ in.

4. 0.65 mi = _____ km

5. 2.4 yd = _____ m

6. 4.25 km = _____ yd

7. 125 ft = _____ km

8. 0.002 m = _____ in.

9. 483 mm = _____ yd

10. 35,140 mm = _____ mi

Problem Solving

Answer the questions.

11. You have decided to enter a 5-kilometer fun run. How many miles is this race?

12. Driving directions tell you that the distance from Toronto to Quebec City is 793 kilometers. How many miles will you drive?

13. You want to build a skateboard ramp that will be 5.1 meters long. How long will the ramp be in feet?

14. You are researching the fuel economy of a popular model of sport utility vehicle. You discover that this vehicle gets 8.9 kilometers per liter. What is this vehicle's fuel economy in miles per gallon?

15. Mai needs nails that are 1.5 inches long to build a bookshelf. How long are the nails in centimeters?

16. A two-person kayak is 15 feet long and 34 inches wide. What is the length and width of the kayak in meters?

LESSON **6** Scale Models and Maps

Scale is a ratio used to change lengths from one unit to another. For example, when you make a model sailboat that has a scale of 1 in. = 20 ft, that means that for every inch of the model, the real sailboat is 20 feet long. In the same way, maps use scale to represent real distances.

> 1 in. model sailboat
> 20 ft real sailboat

Scale: 1 in. = 20 ft

SAIL BOAT

Example

You have a model boat that is 5.5 inches long. The scale on the box reads 1 in. = 20 ft. How long is the real boat?

STEP 1 Write the scale as a ratio.

> $\dfrac{1 \text{ in.}}{20 \text{ ft}}$ $\dfrac{\text{model boat}}{\text{real boat}}$

STEP 2 Set up a proportion.
Let *x* represent the length of the real boat. Write the units in the same order.

> $\dfrac{1 \text{ in.}}{20 \text{ ft}} = \dfrac{5.5 \text{ in.}}{x}$

STEP 3 Multiply.

The real boat is 110 ft long.

> $\dfrac{1}{20} = \dfrac{5.5}{x}$
> $x = 5.5\,(20)$
> $x = 110 \text{ ft}$

ON YOUR OWN

Your friend said the ride from the airport to the stadium in Athens is 34 kilometers. How far is that distance on your map, where the scale is 1 cm = 0.5 km?

Practice

Building Skills

Solve.

1. How long is a truck if a $\frac{1}{40}$ scale model of it is 15 cm long? ($\frac{1}{40}$ means 1 cm of the model = 40 cm in real life)

2. How many centimeters would represent 50 kilometers if you were using a scale of 1 cm = 0.5 km?

3. What distance on the ground does 5 inches on a map represent if the scale is 1 in. = 1.5 mi?

4. How many inches would represent 150 miles if the scale is 1 in. = 10 mi?

5. How long would a $\frac{1}{20}$ scale model of a 40-foot-long airplane be?

6. You are asked to make a $\frac{1}{10}$ scale drawing. How long should a 25-foot-long classroom be in your drawing?

7. How many centimeters would represent 40 kilometers if you were using a scale of 1 cm = 1 m?

8. How many inches would represent 30 feet if you were using a scale of 1 in. = 1 yd?

9. How many inches would represent 12 miles if you were using a scale of 1 in. = $\frac{1}{4}$ mi?

10. If you are using a scale of 1 in. = 0.5 mi, how far does 16.5 inches represent in the real world?

Problem Solving

Solve.

11. You make a drawing of your room using a scale of 2 cm = 1 m. How long should your 2-meter-long bed be in the drawing?

12. Your radio-controlled race car is a $\frac{1}{15}$ scale model. How big is it if the real car is 12 feet long?

13. A map has a scale of 1 cm = 1 km. If two points are 6 centimeters apart on the map, how far apart are they on the ground?

14. Your younger sister's dollhouse is a $\frac{1}{10}$ scale model of a 29-foot-high home. How tall is the dollhouse in inches?

Name _____ Date _____

LESSON ⑦ Using the Distance Formula

Miles per hour or mph is a **rate** telling you how far, or the distance, and how much time it takes to go that far. For example, a car going 60 mph will travel 60 miles in one hour. One way to say this is that distance traveled equals 60 miles × one hour or $d = r \times t$.

d, r, and t are **variables.** Variables are quantities that can change, like your weight. We use letters to stand for variables. On tests, if you know two of the three parts of the formula you can solve for the third part.

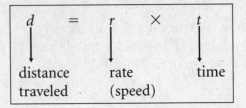

Example

Your class is planning a trip to New York City. You and your friends are planning to walk across the Verrazano Narrows Bridge, which is 13,700 feet long. You know you can walk as fast as 265 feet per minute. How long will it take you to cross the bridge?

STEP 1 Identify your variables.
distance (d) = 13,700 feet
rate (r) = 265 feet per minute
time (t) = what you need to find

STEP 2 Write the formula.
$d = r \times t$

STEP 3 Replace the variables with numbers and solve.
$d = r \times t$
$13,700 = 265 \times t$

$$\frac{13,700}{265} = \frac{265\,t}{265}$$
$$51.7 = t$$

It will take 51.7 or about 52 minutes to walk across the bridge.

ON YOUR OWN

How far would you travel if you drove your car at 55 miles per hour for 3.5 hours?

Practice

Building Skills

Find the distance.

1. rate: 45 miles per hour; time: 1.5 hours

2. rate: 6 meters per second; time: 20 seconds

3. rate: 24 feet per minute; time: 7 minutes

4. rate: 75 kilometers per hour; time: 12 hours

Find the rate.

5. distance: 2.5 kilometers; time: 34 minutes

6. distance: 70 miles; time: 6 hours

Find the time.

7. distance: 75 kilometers;
rate: 60 kilometers per hour

8. distance: 600 meters;
rate: 33 meters per second

Problem Solving

Find distance, rate, or time.

9. A cheetah can run at a rate of 25 meters per second. How far can it run in 15 seconds?

10. At a track meet, Chen runs 4 kilometers in 30 minutes. What is his speed (rate)?

11. The moon is approximately 250,000 miles from Earth. It took an Apollo spacecraft about three days to make the trip from Earth to the moon. How far did the craft travel each day?

12. Harris drives his car at a speed of 35 miles per hour. He drives to a town 20 miles away. How long will it take him to make the trip?

Name _____ Date _____

LESSON 8 — Finding Perimeter

There are 90 ft between bases on a major league baseball diamond. If you started walking from home plate and ended back at home plate, you would have walked 90 ft + 90 ft + 90 ft + 90 ft. You would have walked the **perimeter** of the baseball diamond. The perimeter is the distance around the outside of a figure. When you want to find the perimeter of an object with straight sides, you add the lengths of all of the sides.

Example

Find the perimeter of this rectangle.

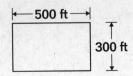

STEP 1 Identify the lengths of all the sides.
In a rectangle, opposite sides are equal.
300 feet, 500 feet, 300 feet, and 500 feet

STEP 2 Add the lengths.

300 ft + 500 ft + 300 ft + 500 ft = 1,600 ft

The perimeter of the rectangle is 1,600 ft.

ON YOUR OWN

Find the perimeter of this triangle.

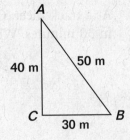

Practice

Building Skills

Find the perimeter.

1.
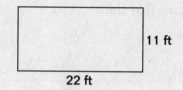

2.
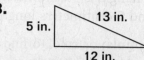

3.
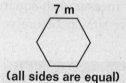

4.
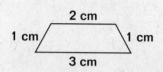

5.
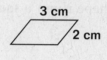

6.
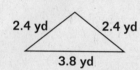

(all sides are equal)

7.
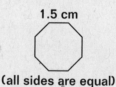

8.
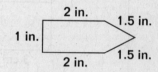

(all sides are equal)

9.

Problem Solving

Find the perimeter.

10. A stop sign has 8 equal 10-inch sides. What is its perimeter?

11. What is the perimeter of a baseball diamond that has 4 equal sides, each 90 feet long?

12. Each side of the square base of the Great Pyramid of Giza, in Egypt, is 230 meters. What is the perimeter of the Great Pyramid?

13. You decide to place weather stripping around your front door. The door is 7 feet tall and 4.5 feet wide. How much weather stripping will you need?

Name _____   Date _____

LESSON **9** **Finding Area of a Rectangle**

Area is the measure of the space inside a flat shape. The area of a rectangle is easy to find if you know how long and how wide it is. You can decide which side you want to call the length; the side next to it will be the width.

This rectangle is 3 centimeters wide and 5 centimeters long.

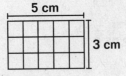

The **formula,** or rule, to find the area of any rectangle, is:

area = length × width, or

$A = l \times w$

Area is measured in square units even if the shape is not a square. Be sure that the units are similar before you multiply.

Example

Find the area of this rectangle.

> 4.5 cm
> 3 cm

STEP 1 Identify the length and width.
The length is 3 centimeters, and the width is 4.5 centimeters.

STEP 2 Write the formula and solve.

$A = l \times w$

$A = 3 \text{ cm} \times 4.5 \text{ cm}$

$A = 13.5 \text{ square cm, or } 13.5 \text{ cm}^2$

ON YOUR OWN

Find the area of this rectangle.

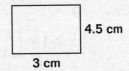

5 cm

2 cm

Name _____ Date _____

Practice

Building Skills

Find the area.

1.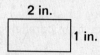
2 in.
1 in.

2.
1 in.
3 in.

3.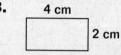
4 cm
2 cm

4.
5.5 cm
1.5 cm

5.
$\frac{1}{2}$ in.
$\frac{1}{4}$ in.

6.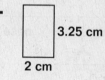
3.25 cm
2 cm

7.
2 ft
2 ft

8.
3 m
3 m

9.
1.5 ft
8.5 ft

10.
12 mi
12 mi

11.
27.3 km
2,000 m

Problem Solving

Find the area.

12. What is the area of an eight-by-ten-inch photograph?

13. The area in front of the school needs new grass. It measures 120 yards long and $53\frac{1}{3}$ yards wide. How many square yards of grass are needed?

14. Central Park, in the heart of New York City, is 2.5 miles long and 0.5 miles wide. What is its area?

15. A wall is 7 feet by 15.5 feet. You have 100 square feet of wallpaper. Do you have enough? Explain.

LESSON 10 — Finding Area of Parallelograms and Triangles

You can use what you learned about rectangles to find area of a parallelogram. You can cut it up and move the pieces around to make a rectangle.

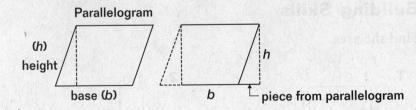

Parallelogram

(*h*) height

base (*b*)

h

b

piece from parallelogram

> area of parallelogram = base × height
> A = bh

To find the area of a triangle you need to know the length of the base and the height.

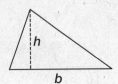

h

b

> area of triangle = $\frac{1}{2}$ × base × height
> A = $\frac{1}{2}$bh

Example

Find the area of this parallelogram.

STEP 1 Write the formula.
The figure is a parallelogram.
A = bh

STEP 2 Identify the base and height.
The base (*b*) is 5 centimeters, and the height (*h*) is 4 centimeters.

STEP 3 Substitute numbers for the letters and multiply.
A = bh
A = 5 cm × 4 cm
A = 20 cm²

The area is 20 square centimeters.

4 cm

5 cm

ON YOUR OWN

Find the area.

1.5 cm

2.5 cm

Name _____ Date _____

Practice

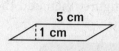

Use the height, not the side measure.

Building Skills

Find the area.

1.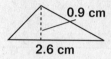
0.9 cm
2.6 cm

2.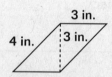
3 in.
4 in. 3 in.

3.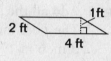
5 cm
1 cm

4.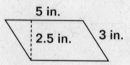
5 in.
2.5 in. 3 in.

5.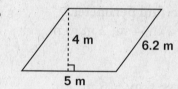
4 m
6.2 m
5 m

6.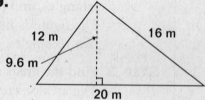
1 ft
2 ft
4 ft

7.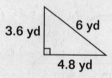
3.6 yd 6 yd
4.8 yd

8.
10 in.
12 in.

9.
12 m 16 m
9.6 m
20 m

Problem Solving

Find the area. (Hint: Draw pictures of the shapes.)

10. The triangular sail on your sailboat is 6 meters across the bottom and 8 meters tall.

11. A diamond-shaped garden measures 20 feet on each side. From one point to the opposite point on the diamond, it measures 30 feet. What is the area of the garden?

12. What is the area of the art room window if it measures 1.2 yards at the bottom and top, is 3.5 yards tall, and has slanted sides that are 4.4 yards long?

13. Which has the greater area: the rectangle or the triangle?

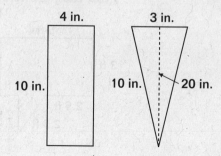
4 in.
10 in.
3 in.
10 in. 20 in.

LESSON ⑪ Area and Perimeter of Complex Figures

You can find the areas and perimeters of all other shapes that are made up of straight sides. To find the perimeter, you add all the lengths of the sides. To find the area, it helps to divide the figure into rectangles, triangles, and parallelograms. You find the area of each smaller figure; then add these areas together.

Example

Find the perimeter and area of the complex figure at the right.

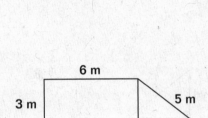

STEP 1 Add all the sides to find perimeter.
The sides measure 5 m, 4 m, 2 m, 6 m, 5 m, 6 m.
The perimeter is
5 m + 4 m + 2 m + 6 m + 5 m + 6 m = 28 m.

STEP 2 Divide the figure into rectangles, triangles, and parallelograms.
The figure is made of one triangle and two rectangles.

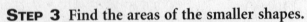

STEP 3 Find the areas of the smaller shapes.
The area of the smaller rectangle:
4 m × 2 m = 8 m^2
The area of the larger rectangle:
6 m × 3 m = 18 m^2
The area of the triangle: $\frac{1}{2}$(4 m × 3 m) = 6 m^2

Step 4 Add all the areas.
8 m^2 + 18 m^2 + 6 m^2 = 32 m^2

The perimeter is 28 m. The area is 32 m^2.

ON YOUR OWN

Find the perimeter and area of this figure.

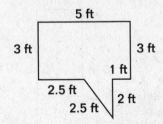

Name _____ Date _____

Practice

Building Skills

The height of a parallelogram may not be one of the sides.

Find the perimeter.

1.
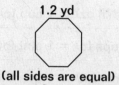
1.2 yd

(all sides are equal)

2.
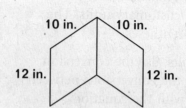
10 in.   10 in.

12 in.   12 in.

3.
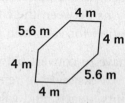
4 m
5.6 m   4 m
4 m   5.6 m
4 m

Find the perimeter and area.

4.
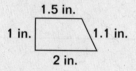
1.5 in.
1 in.   1.1 in.
2 in.

5.
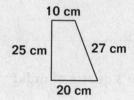
10 cm
25 cm   27 cm
20 cm

6.
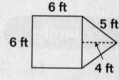
6 ft
5 ft
6 ft
4 ft

Problem Solving

Find the area and perimeter. Draw pictures of the shapes if it will help.

7. In baseball, home plate is a five-sided figure made up of a rectangle 17 inches wide by 8.5 inches long, and a triangle 17 inches wide and 8.5 inches high.

8. The lawn in front of your school is shaped like a parallelogram with parallel side lengths of 15 meters and 25 meters, and a height of 10 meters.

9. You set up four 10-foot-by-3-foot tables to form a capital letter *T*, which is 20 feet wide and 23 feet long. What is the area of the *T*? What is the perimeter?

10. Using the same tables from question 9, you set up the tables again like a wider capital *T*, which is 30 feet wide by 13 feet long. What is the new area and perimeter?

LESSON 12 Customary Units of Capacity

We use customary units of **capacity** every day. A car may need 8 gallons of gasoline to fill its tank. A juice box may contain 6.75 fluid ounces of juice. The chart shows conversions between the common customary units. The abbreviation for each unit is in parentheses.

If you have to convert between units, use the conversion factors. An easy way to remember the conversions between cups, pints, quarts, and gallons is with the equation "**2 × 2 = 4.**" There are **2** cups in one pint, **2** pints in one quart, and **4** quarts in one gallon.

Customary Units Conversion Factors
8 fluid ounces (fl oz) = 1 cup (c)
2 cups (c) = 1 pint (pt)
2 pints (pt) = 1 quart (qt)
4 quarts (qt) = 1 gallon (gal)

Example

How many cups are in 3 quarts of milk?

STEP 1 Find the conversion factor(s).

If you know the relationship between pints and quarts and you know the relationship between pints and cups then you can use these facts to convert from quarts to pints and then from pints to cups.

Quarts → Pints

1 quart (qt) = 2 pints (pt)

Pints → Cups

1 pint (pt) = 2 cups (c)

The conversion factors are:

$\dfrac{2 \text{ pints}}{1 \text{ quart}}$ and $\dfrac{2 \text{ cups}}{1 \text{ pint}}$

STEP 2 Multiply to change units of measure.
You will multiply the number of quarts given by the conversion factor that shows how many pints are in a quart. The quarts units cancel out.

$$\frac{3 \text{ quarts}}{1} \times \frac{2 \text{ pints}}{1 \text{ quart}} = \frac{6 \text{ pints}}{1}$$

STEP 3 Repeat Steps 1 & 2 to convert to the next unit (pints to cups).
You will multiply the number of pints you found in Step 2 by a conversion factor that shows how many cups are in a pint. The pints units cancel out.

$$\frac{6 \text{ pints}}{1} \times \frac{2 \text{ cups}}{1 \text{ pint}} = \frac{12 \text{ cups}}{1}$$

There are 12 cups in 3 quarts.

ON YOUR OWN

How many cups are in 24 fluid ounces?

Name _____ Date _____

Practice

Building Skills

Solve. Show your work.

1. 5 pt = _____ qt

2. 3 gal = _____ qt

3. 16 pt = _____ gal

4. 2.5 qt = _____ c

5. 20 fl oz = _____ c

6. 1 pt = _____ fl oz

7. 28 qt = _____ gal

8. 24 fl oz = _____ pt

9. 1.5 gal = _____ c

10. 160 fl oz = _____ gal

Problem Solving

Solve each capacity problem. Show your work.

11. You need 5 gallons of water to fill a fish tank. How many quarts of water do you need to fill the tank?

12. A load of laundry uses 1 cup of detergent. How many loads of laundry can you do if you buy a 6-quart container of detergent?

13. Your thermos holds 70 fluid ounces of water. How many cups of water does it hold?

14. A restaurant served 8 gallons of salsa last weekend. How many cups of salsa did the restaurant serve?

15. You have enough money to purchase either 14 one-quart containers of milk or 3 one-gallon containers of milk. Which purchase will give you more milk? Explain.

16. You drank 1 gallon of juice and 20 cups of water in one week. Did you drink more juice or water? Explain.

LESSON ⓭ Metric Units of Capacity

In the previous lesson, you learned how to measure capacity using customary units. There are also metric units of capacity. The metric unit for capacity is the liter. The most common metric units of capacity are the liter (L) and the milliliter (mL).

Metric Units Conversion Factors
1 kiloliter (kL) = 1000 liters (L)
1 L = 100 cL
1 cL = 10 mL
1000 mL = 1 L

Remember, when you convert from a larger unit, such as a liter, to a smaller unit, such as a milliliter, you multiply.

- To change L to mL, multiply by 1,000.
- To change mL to L, divide by 1,000.

Example

How many milliliters of water are in the bottle shown?

STEP 1 Identify the units of capacity.

liters (L) milliliters (mL)

STEP 2 Set up the problem.
Because you are changing L to mL, you multiply.

1.24 L × 1,000 mL =

STEP 3 Multiply or divide.

1.24 L × 1,000 mL = 1,240 mL

There are 1,240 mL in 1.24 liters.

Bubbling Springs
WATER
1.24 Liters

(ON YOUR OWN)

How many liters are in 450 milliliters?

Practice

Building Skills

Convert each unit.

1. 2,000 mL = _____ L

2. 4,700 L = _____ mL

3. 5,300 mL = _____ L

4. 400 mL = _____ L

5. 537 L = _____ mL

6. 49.4 L = _____ mL

7. 23,000 mL = _____ L

8. 57,890 mL = _____ L

9. 12 L = _____ mL

10. 20,000 mL = _____ L

Problem Solving

Convert each unit.

11. How many milliliters of water does it take to fill a 2-liter bottle?

12. How many liters does a 500-milliliter jar hold?

13. Alisha bought a six-pack of water. If one bottle of water holds 355 milliliters, how many liters of water are in six bottles?

14. Michael left a measuring cup out in the rain. After the rain stopped, Michael looked at the cup and saw that there were 325 milliliters of water in the cup. How many liters of water were in the cup?

15. Peyton bought a large water bottle that holds 0.8 liters of a beverage. How many milliliters of beverage does it hold?

16. In the chemistry lab, a beaker holds 500 milliliters of solution. How many liters of solution does it hold?

LESSON 14 Converting Units of Capacity

Look at the label on a bottle of juice. You will see that the amount of juice is given in both fluid ounces and in liters. Suppose you have three 1-liter bottles of juice. How many gallons of juice do you have?

To convert from one system of measurement to another, you use a conversion factor that tells you the relationship between the units in each system. For example, 3.8 liters = 1 gallon.

APPLE JUICE

1 Liter • 33.8 fluid ounces

Example

How many liters are in 2 gallons?

STEP 1 Write the number you need to change as a fraction over 1.

$$\frac{2\ gal}{1}$$

STEP 2 Find the conversion factor.

$$\frac{2\ gal}{1} \times \frac{3.8\ L}{1\ gal}$$

STEP 3 Multiply.
So, 7.6 liters equals 2 gallons.

$$\frac{2\ gal}{1} \times \frac{3.8\ L}{1\ gal} = 2 \times 3.8\ L = 7.6\ L$$

ON YOUR OWN

Convert 24 gallons to liters.

Hint: Use the chart on page 106 at the back of this book to find your conversion factor.

Practice

Create your conversion factor so that the unit you wish to change is the top number and the unit you start with is the bottom number.

Building Skills

Convert each unit. Round any decimal to the nearest 100th.

1. 15 mL = _____ tsp.

2. 300 mL = _____ gal

3. 1.5 c = _____ L

4. 3.4 pt = _____ mL

5. 18.7 L = _____ gal

6. 1,024 gal = _____ mL

7. 16 tsp. = _____ mL

8. 9 c = _____ L

9. $\frac{8}{9}$ L = _____ pt

10. $\frac{5}{7}$ qt = _____ L

Problem Solving

Use volume conversions to solve these problems.

11. Your gas tank can hold 12 gallons. How many liters does it take to fill your empty tank?

12. Your measuring cup reads 80 milliliters but the customary marks have rubbed off. How many cups do you have?

13. The gas station has a tank that holds 8,000 liters. How many gallons does it hold?

14. You have a 2-liter bottle and your friend has a $\frac{1}{2}$-gallon bottle. Which bottle can hold more liquid? Explain.

15. A pot holds 2.5 quarts. How many milliliters does it hold?

16. A can of orange juice holds 400 mL. Is this greater than or less than half a liter?

Name _____ Date _____

LESSON ⓵⑤ Customary Units of Weight

Weight is a measure of how heavy an object is. For example, every time you weigh yourself, you are using pounds. Your car may weigh 2 tons. A CD weighs only ounces.

Conversion for Customary Units of Weight
16 ounces (oz) = 1 pound (lb)
2,000 pounds (lb) = 1 ton

There is no abbreviation for ton.

- To change a larger unit to a smaller one, you multiply.
- To change a smaller unit to a larger one, you divide.

In this lesson, you will be using ounce as a measure of weight.

Example

Juan bought a 3-pound bag of sugar. How many ounces of sugar did he buy?

STEP 1 Identify the units of weight.
pounds   ounces

STEP 2 Identify the larger and smaller units.
pounds → ounces
larger → smaller

STEP 3 Multiply or divide.
Because you are changing pounds (larger) to ounces (smaller), you multiply.

3 lb × 16 oz = 48 oz

Juan has 48 oz of sugar.

ON YOUR OWN

Ellen needs 2,800 pounds of topsoil to finish making her garden. How many tons of soil will she need?

Practice

Building Skills

Convert each unit.

1. 3 tons = _____ lb   **2.** 64 oz = _____ lb

3. 2 lb = _____ oz   **4.** 9,000 lb = _____ tons

5. 10 lb = _____ oz   **6.** 128 oz = _____ lb

7. 24 oz = _____ lb   **8.** 3,000 lb = _____ tons

9. 4 tons = _____ lb   **10.** 0.5 tons = _____ oz

Problem Solving

Solve.

11. Amber and Maria are packing a cooler for a picnic, and they need more ice. Maria goes to the store and buys a 5-pound bag of ice. How many ounces of ice did she buy?

12. Many restaurants serve quarter-pound hamburgers. How many ounces of meat are in 1 quarter-pound hamburger? (*Hint:* 1 quarter-pound = $\frac{1}{4}$ pound = 0.25 pound)

13. A team of marine biologists is working to help a beached whale move back out to sea. If the whale weighs 1.2 tons, how many pounds does it weigh?

14. The owner's manual says that your car weighs 5,000 pounds. How many tons does your car weigh?

15. The grocery store is selling a 12-ounce box of cereal for $2.40, and a 1-pound box of cereal for $3.20. Which box of cereal has the lower price per ounce? Explain.

16. A crane can lift a steel beam that weighs 6.5 tons. How many pounds does the beam weigh?

LESSON 16 Metric Units of Mass

Mass is the amount of matter in an object. Metric units of mass are measured in milligrams, grams, and kilograms. We usually use milligrams to measure medicine. A dollar bill weighs about a gram. Two dozen eggs weigh about one kilogram.

Conversion for Metric Units of Mass
1,000 milligrams (mg) = 1 gram (g)
1,000 grams (g) = 1 kilogram (kg)
1,000 kilograms (kg) = 1 metric ton

- To change from a larger to a smaller unit, you multiply.
- To change from a smaller to a larger unit, you divide.

Example

What is the mass of the bananas in kilograms?

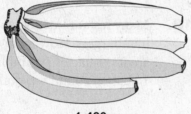

1,400 g

STEP 1 Identify the units.
grams and kilograms.

STEP 2 Identify the smaller and larger units.
grams → kilograms
smaller → larger

STEP 3 Multiply or divide.
Because you are changing grams (smaller) to kilograms (larger), you divide.
1,400 g ÷ 1,000 kg = 1.4 kg

The bananas weigh 1.4 kg.

ON YOUR OWN

How many milligrams are in 24 grams?

Practice

Building Skills

> Larger to smaller, multiply.

Convert each unit.

1. 1.5 kg = _____ g

2. 4,500 g = _____ kg

3. 1,250 mg = _____ g

4. 2.5 kg = _____ g

5. 6 g = _____ mg

6. 3 kg = _____ g

7. 500 mg = _____ g

8. 8,000 g = _____ kg

9. 1.4 kg = _____ g

10. 0.5 g = _____ mg

Problem Solving

Use metric units of mass to solve these problems. Show your work.

11. For a science project, your class needs a total of 2 kilograms of salt, but the scale only measures mass in grams. How many grams of salt does your class need?

12. A fisherman caught a rainbow trout with a mass of 527 grams. What is the mass of the fish in kilograms?

13. An adult must take 60 milligrams of vitamin C per day. How many grams of vitamin C should an adult take a day?

14. An emperor penguin can lay an egg with a mass of up to 0.68 kilograms. What is this mass in grams?

15. The grocery store is selling coffee in 750 g bags. If you purchase four of these bags, how many kilograms of ground coffee will you have?

16. The label on a bag of 10 pretzels says that the pretzels contain 250 milligrams of sodium per serving. A serving is equal to one bag. How many grams of sodium are in the bag?

Name _____   Date _____

LESSON 17 Converting Units of Mass and Weight

Mass and weight are both measures of how heavy an object is. We can use the conversions below to convert between customary units and metric units.

1 gram (g) = 0.035 ounce (oz)

1 kilogram (kg) = 2.2 pounds (lb)

Example

How many pounds are in 500 grams?

STEP 1 Find the conversion factors.
If you know the relationship between pounds and kilograms, and you know the relationship between grams and kilograms, you can use these facts to convert from grams to kilograms and from kilograms to pounds.

Grams → Kilograms   Kilograms → Pounds
1,000 grams (g) = 1 kilogram (kg)   1 kilogram (kg) = 2.2 pounds (lb)

The conversion factors are:

$$\frac{1 \text{ kg}}{1000 \text{ g}} \qquad \frac{2.2 \text{ lb}}{1 \text{ kg}}$$

STEP 2 Multiply to change units of measure.
You will multiply the number of grams given by the conversion factor that shows how many grams are in a kilogram. The gram units cancel out.

$$\frac{500 \text{ g}}{1} \times \frac{1 \text{ kg}}{1000 \text{ g}} = \frac{0.5 \text{ kg}}{1}$$

STEP 3 Repeat Steps 1 and 2 to convert to the next unit.
You will multiply the number of kilograms you found in Step 2 by a conversion factor that shows how many pounds are in a kilogram. The kilogram units cancel out.

$$\frac{0.5 \text{ kg}}{1} \times \frac{2.2 \text{ lb}}{1 \text{ kg}} = \frac{1.1 \text{ lb}}{1}$$

There are 1.1 lb in 500 g.

(ON YOUR OWN)

How many kilograms are there in 1 ton?

Practice

Building Skills

Convert each unit.

1. 7 oz = _____ g

2. 70 oz = _____ g

3. 11 lb = _____ kg

4. 100 oz = _____ g

5. 2 tons = _____ kg

6. 16 kg = _____ lb

7. 6.6 lb = _____ g

8. 450 g = _____ lb

9. 100 g = _____ oz

10. 32 oz = _____ kg

Problem Solving

Solve.

11. You have a bag of granola with a mass of 60 grams. How many ounces of granola do you have?

12. If you buy a 5-kilogram bag of potatoes, how many pounds of potatoes do you have?

13. A mason used 5,280 pounds of cement to build a walkway. How many kilograms of cement did the mason use?

14. Sam has a pet bird that weighs 4 ounces. How many grams does the bird weigh?

15. Mr. Jones bought a 50-pound bag of sand to put in his daughter's sandbox. How many kilograms of sand did he buy?

16. A truck has a mass of 6,000 kilograms. Will the truck be able to cross a bridge with a weight limit of 12,500 pounds? Explain.

Name _____ Date _____

LESSON 18 Points and Lines

Points, lines, and angles are the building blocks of geometry. The tiles that make up this floor are repeated copies of the same simple shapes.

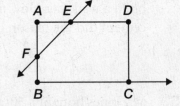

Art, nature, and the shapes that surround us every day are made up of points, lines, and angles.

A **point** is an exact location. It has no size, shape, or direction.

A **line segment** has 2 endpoints. An endpoint names where a line starts and stops. Use endpoints to name a line segment.

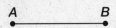

A **line** is a straight path that extends without end in opposite directions. Use 2 points on a line to name a line.

A **plane** is like the flat surface of this page. But like a line, a plane goes on forever. Lines and points can be on a plane.

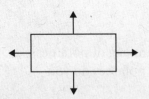

A **ray** has one endpoint and then goes on without stopping in one direction only. Use the endpoint as the first letter when naming a ray.

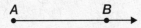

Example

Look at the drawing. How many of the geometric parts can you identify?

STEP 1 How many points are there?
Remember, a point marks a position but has no size, shape, or direction. There are 6 points.

STEP 2 How many segments are there?
A segment is a part of a line. It has a point where it starts and stops. There are 9 segments in the figure.

STEP 3 What else is in the drawing?
6 rays and 1 line.

ON YOUR OWN

Look at this figure. How many geometric parts can you identify?

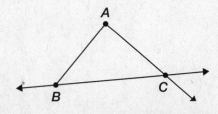

Name _____ Date _____

Practice

Building Skills

Find the number of parts.

1. How many line segments are in this figure?

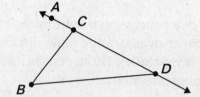

2. How many points does the figure have?

Use this figure for questions 3 and 4.

3. How many line segments make up the sides of this figure?

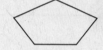

4. How many points make up the corners of this figure?

Use this figure for questions 5 and 6.

5. How many rays are in the figure?

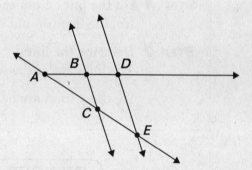

6. How many lines are in the figure?

Use this 2-dimensional figure for questions 7 and 8.

7. How many points are at the corners of this figure?

8. How many line segments make up the figure?

Problem Solving

Use the figure on the right to answer numbers 9–12.

9. Name 3 different line segments.

10. Name 4 different rays.

11. Give 6 names for this line.

12. Give another name for the ray *RQ*.

13. Name at least 4 parts of the geometric figure on the right.

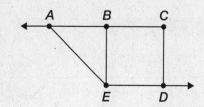

Name _____   Date _____

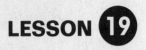

LESSON 19 Parallel and Perpendicular Lines

Sometimes lines are drawn in special ways. **Parallel** lines are in the same plane. But parallel lines never meet and have no common points. Train tracks are examples of parallel lines. **Perpendicular** lines **intersect,** or cross, to form square corners.

Think of two pencils as straight lines. If the pencils can lie on the same piece of paper or plane, the same distance apart, and are *not* intersecting, then they are **parallel.** If the pencils intersect and form 4 corners, then they are **perpendicular.**

You can identify lines as parallel (||), perpendicular(⊥), or intersecting (lines that cross each other, but are not perpendicular).

Example

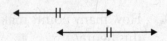

Are the lines shown parallel, perpendicular, or intersecting (and not perpendicular)?

STEP 1 Do the lines cross each other?
No, the lines do not cross, so they are not perpendicular or intersecting.

STEP 2 Describe the lines.
The lines are in the same plane and are an equal distance apart.

So, these lines are parallel.

(**ON YOUR OWN**)

Are the lines shown parallel, perpendicular, or intersecting (and not perpendicular)?

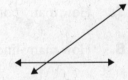

Practice

Building Skills

Identify the pairs of lines as parallel, perpendicular, or intersecting but not perpendicular. Use this figure for questions 1–10.

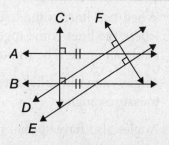

1. *A* is _____ to *D*.

2. *A* is _____ to *C*.

3. *D* is _____ to *E*.

4. *C* is _____ to *D*.

5. *B* is _____ to *A*.

6. *D* is _____ to *B*.

7. *C* is _____ to *F*.

8. *D* is _____ to *F*.

9. *A* is _____ to *F*.

10. *E* is _____ to *F*.

Problem Solving

Describe how the lines are related.

11. The yard lines on a football field

12. The edges of a poster that form a corner

13. Draw a line through the center of a pyramid that is perpendicular to the ground. Trace a line along one edge of the base of the pyramid.

14. The white line on the edge of a road and a post that holds a stop sign

15. The rungs of a ladder

16. The strings of a guitar

LESSON 20   Angles

When two lines come together at the same point, they form an **angle** (∠). The point where the lines come together is called the **vertex** of the angle. Three letters (points) name an angle. The middle letter is the vertex. The angle shown is ∠*ABC* or ∠*B*.

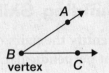

Angles come in different sizes and are measured in degrees (°). A **protractor** measures angles.

Angles also have special names based on their size or measure. The table below includes the names of the different types of angles.

Angle Name	Measure	Hints and Facts
acute	Between 0° and 90°	Smaller than a right angle
right	90°	There's a small square in the corner. Squares have right angles.
obtuse	Between 90° and 180°	Bigger than a right angle
straight	180°	Lines form straight angles.

Example

What type of angle is shown?

STEP 1 Is the angle a right angle or a straight angle?
It is neither a right angle nor a straight angle.

STEP 2 Compare it to a straight angle.
It is smaller than a straight angle, so it is either obtuse or acute.

STEP 3 Compare it to a right angle.
The angle is smaller than a right angle.

So, it is acute.

(ON YOUR OWN)

What type of angle is shown?

Practice

Building Skills

Name each type of angle.

1.

2.

3.

4.

5.

6.

A

7.

A
B   C

8.

P   R
S

9.

M
N
P

Problem Solving

Solve.

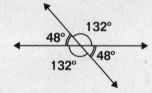

48°   132°
132°   48°

Use this figure for questions 10–12.

10. What are the measures of the acute angles?

11. What are the measures of the obtuse angles?

12. If you add a 48° angle to a 132° angle, what type of angle have you made?

13. The pointed end of a pizza slice measures 30°. How many of these slices would make up the entire pizza? [Hint: a circle is 360°.]

14. A skateboard ramp measures 45° up from the ground. What is the measure of the angle between the ramp and the ground in front of it?

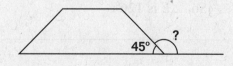

45°   ?

LESSON 21 Pair of Angles

Sometimes one angle relates in a special way to another angle. For example, some pairs of angles are always equal to each other. Some pairs add to 180° or 90°.

Corresponding Angles

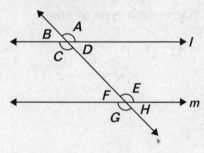

If you have a pair of parallel lines and a line that intersects them, the lines form 8 angles: ∠A, ∠B, ∠C, ∠D, ∠E, ∠F, ∠G, ∠H. The pairs of angles created by the intersecting line have special relationships with each other. When angles are formed by a line intersecting two parallel lines, the **corresponding angles** are in the same position along a different parallel (‖) line. Corresponding angles are equal in measure.

∠A = ∠E = ∠C = ∠G

∠B = ∠F = ∠D = ∠H

Vertical Angles
When two lines intersect, the opposite angles are **congruent,** equal in size.

Complementary Angles
Two angles whose sum equals 90° are **complementary.**

Supplementary Angles
Two angles whose sum equals 180° are **supplementary.**

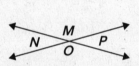

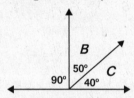

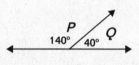

∠M = ∠O   ∠N = ∠P

∠B + ∠C = 90°

∠P + ∠Q = 180°

Example

If ∠A in the diagram at the top of the page is 126°, what is the measure of ∠E ?

STEP 1 What is the position of the pair of angles?
∠A and ∠E are in the same position (top right) in the intersections of each of the parallel lines, l and m.

STEP 2 Classify the angles.
∠A and ∠E are corresponding angles.

STEP 3 What is the relationship between the angles?
Corresponding angles are equal.

So, ∠E is 126°.

(**ON YOUR OWN**)

Find the measure of an angle that is supplementary to a 126° angle.
[Hint: Supplementary angles add to 180°.]

Practice

Building Skills

Find the measure of each angle.

Use this figure for questions 1–10.

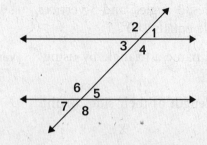

1. If ∠1 is 42°, what is the measure of ∠7?

2. If ∠5 is 50°, what is the measure of ∠8?

3. If ∠2 is 150°, what is the measure of ∠4?

4. If ∠6 is 124°, what is the measure of ∠5?

5. If ∠8 is 140°, what is the measure of ∠4?

6. If ∠2 is 112°, what is the measure of ∠6?

7. If ∠2 is 112°, what is the measure of ∠8?

8. If ∠1 is 25°, what is the measure of ∠5?

9. If ∠1 is 35°, what is the measure of ∠8?

10. If ∠4 is 146°, what is the measure of ∠6?

Problem Solving

Solve.

11. You try to parallel park, and the angle that your tire forms with the curb is 27°. What obtuse angle does the other side of the curb form with your tire?

12. While riding your motorcycle, you brake hard to miss hitting a dog. Your tire leaves a skid mark as you come to a stop. The skid mark forms four angles when it crosses the painted line on the side of the road. One angle formed is 62°. What other angles are formed?

Name _____ Date _____

LESSON ⟨22⟩ Properties of Triangles

A **triangle** is a **polygon** with 3 sides, 3 angles, and 3 vertices. The sum of the angles in a triangle equals 180°.

Triangles have 3-letter names. You name a triangle by using the letters of the vertices.

Triangles can be classified by the length of their sides or the size of their angles.

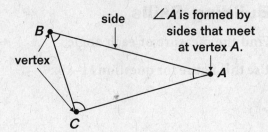

side
∠A is formed by sides that meet at vertex A.

vertex

A scalene triangle is a triangle whose sides all have different lengths.

An isosceles triangle is a triangle that has two sides of the same length.

An equilateral triangle is a triangle whose sides are all the same length.

An acute triangle is a triangle whose angles all are acute.

A right triangle is a triangle with one right angle.

An obtuse triangle is a triangle with one obtuse angle.

Example

Find the missing angle measure in the triangle.

STEP 1 Add the two given angles.
angle *QPR* (90°) + angle *PQR* (40°)
90° + 40° = 130°

STEP 2 Subtract this answer from 180°.
180° − 130° = 50°

Angle *PRQ*, the third angle, is 50°.

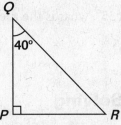

ON YOUR OWN

Find the third angle in this triangle.

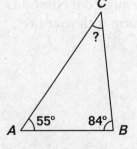

Name _____ Date _____

Practice

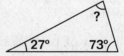

All triangles add up to 180°.

Building Skills

Classify each triangle as an acute, obtuse, or right triangle. Then, find the measure of the missing angle in each.

1.
52°
64° ?

2. ?
27° 73°

3. ? 126° 25°

Problem Solving

Use triangle properties to solve these problems.

4. You are building shapes from colored rods with your nephew. You have a 2-in. rod and 2 5-in. rods. If you make a triangle, what type of triangle will you make?

5. You go for a run on three roads that form a triangle. You know that the first and second roads are 0.8 mi and 1.3 mi long. The third road is 2 miles long. Which type of triangle does this arrangement form?

6. A line supporting a volleyball net breaks. You reattach a rope to the pole at a 55° angle and place the stake in the ground. The pole now stands perpendicular to the ground. What angle does the rope make with the ground?

7. Because each triangle has angles that add to 180°, how many degrees do the angles in this hexagon add up to? (Hint: Count the number of triangles in the hexagon.)

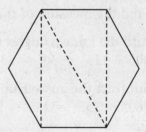

8. A chair's back, arm, and seat forms a triangle. The angle formed between the seat and arm measures 35°. The angle formed by the seat and back measures 77°. What is the measure of the angle formed by the back and arm?

9. Draw an acute scalene triangle.

LESSON 23 Properties of Quadrilaterals

All **quadrilaterals** have four sides, four angles, and four vertices. No matter what the shape of the quadrilateral, the four angles always add to 360°.

Rectangles and squares are common quadrilaterals.

The relationships between sides and angles give each special type of quadrilateral its specific name and properties.

A **trapezoid** is a quadrilateral with only one pair of parallel sides. A **parallelogram** is a quadrilateral with two pairs of parallel sides. Parallelograms also have their opposite sides equal and opposite angles equal.

A **rectangle** has four right angles, and its opposite sides are equal.

A **rhombus** has four equal sides that are parallel.

A **square** has four equal sides and four right angles.

Kites have two adjoining sides equal at top and bottom.

Understanding these properties can be helpful when you need to find certain answers about these shapes.

On tests, you may be asked to find missing angles. With quadrilaterals, you just subtract the sum of the three other angles from 360°.

Trapezoid

Parallelogram

Rectangle, a type of parallelogram

Rhombus, a type of parallelogram

Square, a type of rectangle

Kite

Example

Find the measure of the missing angle.

STEP 1 Add the known angles together.
90° + 115° + 30° = 235°

STEP 2 Subtract the answer in Step 1 from 360°.
360° − 235° = 125°

The missing angle is 125°.

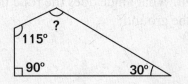

ON YOUR OWN

Find the measure of the missing angle.

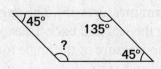

Practice

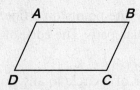

Building Skills

Use this figure for questions 1–4.

1. In parallelogram *ABCD*, if ∠*A* = 155°, what is the measure of ∠*B*?

2. In parallelogram *ABCD*, if ∠*D* = 30°, what is the measure of ∠*A*?

3. In parallelogram *ABCD*, if $\overline{AB}$ is 10 cm long, and $\overline{BC}$ is 5 cm long, how long is $\overline{CD}$?

4. In parallelogram *ABCD*, if $\overline{AB}$ is 6-in. long, and $\overline{DA}$ is 2-in. long, how long is $\overline{BC}$?

Problem Solving

Solve.

5. Your neighborhood swimming pool is a rectangle that is 9 ft wide and 15 ft long. How far is it around the whole pool?

6. A flag has a parallelogram on it that contains a 45° angle. What are the measures of the other 3 angles?

7. Every baseball diamond has 4 equal distances between bases and 90° angles connecting the bases. What type of shape connects the 3 bases and home plate?

8. Two angles of a quadrilateral are 25° and 35°. The third angle is twice the measure of the fourth angle. What are the measures of the third and fourth angles?

LESSON 24 Properties of a Circle

A circle is a closed set of infinite points on a line. Picture a sprinkler watering in a circular path. The edge of the circle is the **circumference.**

The distance from the center of a circle to any point on the edge is called the **radius.** A line segment whose endpoints lie on the circle is called a **chord.** A chord that goes through the center of the circle is called a **diameter.** Notice that the diameter is always twice as long as the radius. Every circle = 360°.

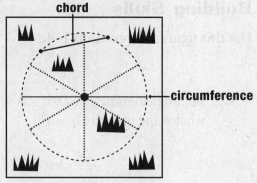

radius + radius = diameter

Example

If $\overline{AB}$ is 20 in., what is the measurement of $\overline{AO}$?

STEP 1 Define the lines.
$\overline{AO}$ is the radius. $\overline{AB}$ is the diameter.

STEP 2 Set up the problem.
Since a radius is half the diameter, divide 20 by 2.
$20 \div 2 = 10$

$\overline{AO}$ is 10 in.

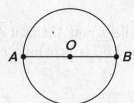

ON YOUR OWN

If $\overline{AB}$ is 6.5 in. what is the measure of $\overline{AC}$?

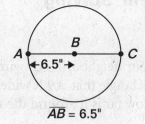

$\overline{AB} = 6.5"$

Practice

Building Skills

Solve.

Use these figures for questions 1–8.

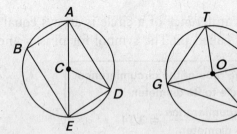

1. Name one radius in circle C.

2. Name two radii of circle O.

3. Name a line segment in circle C that is not a chord.

4. Name the sides of inscribed ∠B.

5. How many chords are shown in circle O?

6. What are the two shortest chords in circle C?

7. $\overline{SG}$ is a diameter. What is the sum of the central angles ∠SOJ and ∠GOJ?

8. $\overline{GS}$ is a diameter of circle O, and ∠TOS = 86°. What is the measure of ∠TOG?

Problem Solving

Solve.

Use the figures above for questions 9 and 10.

9. In a triangle, angles that are opposite from equal sides are equal. Use this fact to help find the measure of ∠OJG if ∠GOJ is 150°.

10. If ∠TOS = 86°, find the measure of ∠OTS.

11. You are buying wheel covers for your car. The diameter of your wheels is 28 in. Wheel covers are sold by radius size. What is the correct radius size for your wheels?

12. You have 3 points on a circle and you connect them. What is the sum of the 3 inscribed angles you formed? [Hint: Think about the sum of the angles in a triangle]

13. On the exercise bikes in the gym, the speedometer is a circle. Every 10° of the central angle is 0.5 miles per hour (mph). What is the measure of the central angle between 0 mph and 5.5 mph?

14. You divide a meat pie into 5 equal pieces. What is the central angle that each piece forms?

Name _____ Date _____

LESSON 25   Circumference of a Circle

Once you can find the radius and diameter of a circle, you can also find the distance around the circle.

This distance around the edge of a circle is called the **circumference.**

The circumference of a circle is always equal to the length of the diameter times a number called **pi.** The symbol for pi is π and is equal to about 3.14.

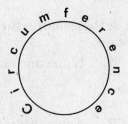

> **pi** is the ratio of the circumference of a circle to its diameter.
>
> $\pi = \dfrac{\text{circumference}}{\text{diameter}} = 3.14$

To find a circumference of a circle, multiply the diameter by π or the radius by 2 times π. The radius is one half the diameter.

> Circumference of a circle:
> $C = 2\pi \times r$ or
> $C = 2\pi r$ or πd

You can also use $\frac{22}{7}$ as the value for pi.

Example

What is the circumference of a circle with a radius of 8 m?

STEP 1 Use the formula for circumference.
$C = 2\pi r$

STEP 2 Put the numbers in the formula and solve.
You know that the radius is 8 m.
$\pi \approx 3.14$
$C = 2 \times 3.14 \times 8 = 50.24$ m

$C = \mathbf{50.24}$ **m**

ON YOUR OWN

What is the circumference of a circle with a radius of 4 cm?

Practice

Building Skills

Find each distance. Use 3.14 for π.

1. What is the circumference of a circle with a diameter of 4 ft?

2. What is the circumference of a circle with a radius of 5 mm?

3. What is the circumference of a circle with a diameter of 19 yd?

4. What is the circumference of a circle with a radius of 2 mi?

5. What is the circumference of a circle with a diameter of 3 cm?

6. What is the circumference of a circle with a radius of 6 in.?

7. What is the circumference of a circle with a radius of 4.5 m?

8. What is the diameter of a circle with a circumference of 18 ft?

Problem Solving

Solve using the formula for circumference. Use 3.14 for π.

9. A bicycle wheel has a diameter of 36 in. How far does it go each time the wheel turns all the way around?

10. A CD has a radius of about 5 cm. How far is it around the edge of the disc?

11. A plastic flying disk has a diameter of 1 ft. What is the circumference?

12. A trainer has a horse on a 16-ft length of rope. The horse walks in a circle around the trainer. How far has the horse walked if it walks around the trainer once?

Name _____ Date _____

LESSON **26** **Area of a Triangle**

Now that you can find areas of rectangles, you can use a similar method to find the area of a triangle.

> **Area = $\frac{1}{2}$ × base × height or $A = \frac{1}{2}bh$**

The base can be any side of the triangle. The height is a line drawn from the top of the triangle and perpendicular to the base or an extension of the base.

In a right triangle, the height is one side of the triangle.

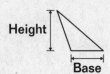
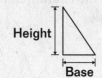
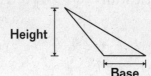

Example

Find the area of the triangle below.

3 cm
4 cm

STEP 1 Write the formula to find the area of a triangle.

$A = \frac{1}{2}bh$

STEP 2 Put the numbers you know into the formula.

$A = \frac{1}{2} \times 3\ \text{cm} \times 4\ \text{cm} = 6\ \text{sq cm or } 6\ \text{cm}^2$

STEP 3 Solve.

The area of the triangle is 6 cm^2.

(**ON YOUR OWN**)

Find the area of this triangle.

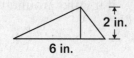

2 in.
6 in.

Practice

Building Skills

Solve.

1. Find the area of the triangle.

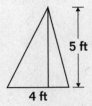

5 ft

4 ft

2. Find the area of the triangle.

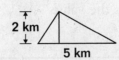

2 km

5 km

3. Find the area of the triangle.

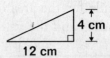

4 cm

12 cm

4. Find the height.

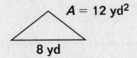

A = 12 yd²

8 yd

5. Find the height.

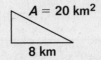

A = 20 km²

8 km

6. Find the base.

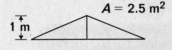

A = 2.5 m²

1 m

Problem Solving

Solve.

7. How many square feet is a triangular sail if the base is 8 ft and the sail is 9 ft high?

8. You start to fold an airplane out of a piece of paper and have a triangle that has a base of 8.5 in. and a height of 11 in. What is the area of this triangle?

9. How much fabric do you need to make a triangular flag that is 1 yd wide and has a base of 1.5 yd?

10. You are painting a house, and one outside wall is a rectangle with a triangle on top of it. What is the area of this wall?

4 m   6 m

8 m

11. What is the area of the traffic arrow on this street?

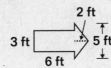

2 ft

3 ft   5 ft

6 ft

12. A triangle-shaped hang glider has an area of 60 sq ft. The wings are 12 ft from tip to tip. What is the distance from the front to the back of the glider?

Name _____ Date _____

LESSON 27 Area of a Circle

You have already learned the formula for the distance
around the edge of a circle, or the circumference.

> **Circumference = 2 × π × r or C = 2πr or πd**

The formula for the area of a circle also uses the value
for the radius and the value of pi (π).

> **Area of a circle = π × radius² or A = πr²** ← **exponent, $r^2 = r \times r$**

In the same way you did with rectangles and triangles, you can use this formula to find
the area if you know the radius, or to find the radius if you know the area.

$C = 2\pi r$ and $A = \pi r^2$

Example

Find the area of the circle. For π use 3.14.

3 cm

STEP 1 Write the formula for the area of a circle.
$A = \pi r^2$

STEP 2 Put the number you know into the formula and solve.
$A = 3.14 \times (3 \text{ cm})^2 = 28.26 \text{ cm}^2$

The area of this circle is 28.26 cm².

ON YOUR OWN

Find the area of the circle.

5 m

Practice

Building Skills

Solve. Use 3.14 for π.

1. Find the radius of the circle.

Area = 12.56 ft²

2. Find the area of the circle.

4 km

3. Find the area of the circle.

20 yd

4. Find the radius of the circle.

A = 113.04 cm²

5. Find the radius of the circle.

A = 78.5 yd²

6. Find the diameter of the circle.

A = 153.86 m²

Problem Solving

Solve. Use 3.14 for π.

7. What is the area of a circular parachute that measures 14 ft across the center?

8. You see an old photo of your grandparents playing with a large circular hoop. If the diameter of the hoop is 4 ft, what is the area inside it?

9. Your class has been asked to paint a big smiley face on the pavement in front of the school stadium. You have enough paint to cover 200 sq ft. To the nearest foot, how big should the radius of the circle be? Use 3.14 for π.

10. Each ring of this archery target is 2 in. wide, and the bull's-eye is 4 in. across. What is the area of the entire target?

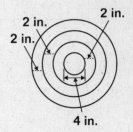

2 in. 2 in. 2 in. 4 in.

LESSON 28 Similar and Congruent Figures

With geometric figures, if two figures are the same, then you say they are **congruent. Congruent figures** have the same shape and same size. If you know that two figures are congruent, then you know that their measurements are the same.

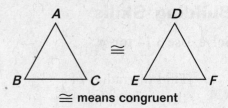

≅ means congruent

If two figures are **similar,** it means that the figures have the same shape and that the corresponding angles are equal.

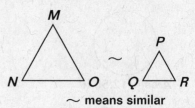

∼ means similar

Figures are similar if they have the same proportions. The rectangles are similar.

$$\frac{3}{5} = \frac{6}{10}$$

Example

Barbara had a 3-in.-by-5-in. photo enlarged to a 6-in.-by-10-in. photo. Is the enlargement similar to the original?

STEP 1 Find the ratio of the lengths. $\frac{5}{10}$

STEP 2 Find the ratio of the widths. $\frac{3}{6}$

STEP 3 Set up a proportion. $\frac{5}{10} \times \frac{3}{6}$
$30 = 30$

Yes, they are similar because they are proportional.

ON YOUR OWN

A regulation high school basketball court is 50 ft × 84 ft. A college court is also 50 ft wide but is 94 ft long. Is the college court similar to the high school court?

Practice

Building Skills

Find the length of each side.

Use this diagram for questions 1–3.
MNOP ≅ *GHIJ*.

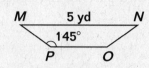
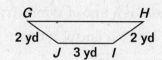

1. $\overline{MP}$

2. $\overline{OP}$

3. $\overline{GH}$

Use this diagram for problems 4–6.
ABCDEF ~ *MNOPQR*.

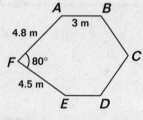
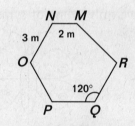

4. $\overline{BC}$

5. $\overline{QR}$

6. $\overline{MR}$

Problem Solving

Solve using the rules of similarity and congruence.

7. You are enlarging a picture and want the two rectangles to be similar. The width went from 1 in. to 12 in. The height was $\frac{2}{3}$ in. How tall will the final picture be?

8. On a certain day, every motorcycle that comes off an assembly line is congruent. If the first motorcycle's rear wheel has a diameter of 22 in., what is the diameter of the rear wheel on the 37th motorcycle?

9. You have a 3-ft-high and 5-ft-long flag. If a huge flag in a parade is similar and 15 ft high, how long is it?

10. Are the areas of congruent figures always the same? Explain.

LESSON 29 Symmetry

Some figures are congruent to each other, and they can also have **symmetry**. A figure has symmetry if it can be folded so that the two parts of the figure match. The fold line across the figure is a **line of symmetry.**

Some figures have no line of symmetry.

No symmetry

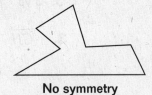

No symmetry

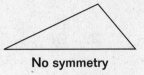

No symmetry

Some figures have only 1 line of symmetry.

1 line of symmetry

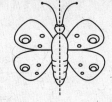

1 line of symmetry

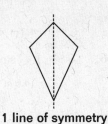

1 line of symmetry

Some figures have more than 1 line of symmetry.

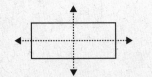

More than 1 line of symmetry

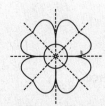

4 lines of symmetry

Example

How many lines of symmetry are in the triangle?

Draw a line to split a figure into congruent halves.
Congruent means equal.

There are 3 lines of symmetry in the triangle.

ON YOUR OWN

How many lines of symmetry are there in a square?

Practice

Building Skills

Solve.

Use this figure to answer questions 1 and 2.

1. How many lines of symmetry does this regular pentagon have?

2. How many new angles are formed at the place where the lines of symmetry cross?

Use this figure to answer questions 3 and 4.

3. How many lines of symmetry does this figure have?

4. Where the lines of symmetry cross, are the angles equal?

Use this figure to answer questions 5 and 6.

5. How many lines of symmetry does this arrow have?

6. What does this arrow look like folded in half along the vertical line of symmetry?

Use this figure to answer questions 7 and 8.

7. How many lines of symmetry does this smiley face have?

8. Does this smiley face look the same for all lines that can cut it in half? Explain.

Problem Solving

Solve using symmetry.

9. Does a yin/yang symbol have symmetry?

10. How many lines of symmetry does a soccer field have?

11. What numerical digits have symmetry?

12. What capital letters have symmetry (not including cursive)?

13. Does a starfish have a line of symmetry? Explain.

14. All regular figures follow a pattern with lines of symmetry. Use this pattern to find the number of lines of symmetry on a regular 15-sided figure.

Name _____ Date _____

LESSON **30** The Pythagorean Theorem

Patterns appear throughout geometry, and one of the most famous is the relationship between the sides in a right triangle. The special formula, known as the **Pythagorean theorem,** helps you find distances between points on a coordinate plane, find missing triangle side lengths, or determine if a triangle has a right angle.

The formula for the Pythagorean theorem is: $a^2 + b^2 = c^2$

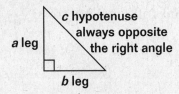

c hypotenuse
always opposite
the right angle
a leg
b leg

If three triangle side measurements work in the formula, then you know you have a right triangle.

Example

Find the length of the third side in this right triangle.

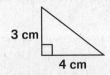

3 cm

4 cm

STEP 1 Write out the formula.
$a^2 + b^2 = c^2$

STEP 2 Put the numbers you know into the formula.
$3^2 + 4^2 = c^2$

STEP 3 Solve.
$3^2 + 4^2 = 9 + 16 = 25$ $c^2 = 25$
$\sqrt{25} = 5$ $c = 5$

The length of the third side is 5 cm.

ON YOUR OWN

A ladder is placed against the side of a building at 24 ft. The ladder is 25 ft tall. How far away from the building is the ladder?

Practice

Building Skills

Solve.

1. Do the sides of a triangle measuring 7 in., 24 in., and 25 in. make up a right triangle?

2. Do the sides 12 cm, 15 cm, and 20 cm make up a right triangle?

3. Find the missing side.

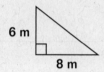
6 m
8 m

4. Find the missing side.

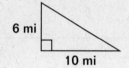
6 mi
10 mi

5. Find the missing side.

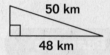
50 km
48 km

Problem Solving

Solve using the Pythagorean theorem.

6. How far is it diagonally from corner to corner of a 50-yd-by-100-yd rectangular field? Hint: Draw a picture.

7. Televisions are measured by the diagonal distance across the screen. Your television screen has a 40-in. diagonal and a height of 25 in. How wide is the screen?

8. A baseball diamond is a square with 90 feet on each side. How far is it between home plate and second base?

9. You have a 20-ft cord on your electric guitar from your speaker in the corner of your garage. If the garage is 17 ft wide and 12 ft long, can the cord reach the opposite corner? Explain.

10. You are setting up a tent for an outdoor party. You want to make sure that the poles are at right angles to the ground. Each pole is 2 m high and the tie ropes are 2.5 m long. How far (in meters) from the poles should you put the stakes to make right triangles?

11. Your science class is trying to design high-flying kites. You use 40 ft of string to fly the kite. Your friend stands under the kite and is 30 ft away from you. How high is your kite flying? (Hint: Draw a picture.)

LESSON 31 Translations

The **coordinate plane** is made up of two number lines that are perpendicular to one another. The horizontal line is the *x*-axis. The vertical line is called the *y*-axis. You can name a point on the coordinate plane by using a pair of numbers, an *x*-coordinate and a *y*-coordinate, (x, y). The point where the *x*-axis and *y*-axis cross is called the **origin,** $(0, 0)$.

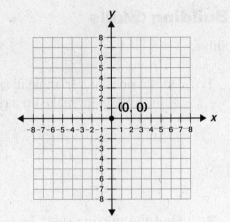

A **translation** is when you *slide* a figure to a new location on a coordinate plane. Every point in the figure slides in the same direction and the same distance. The **slide coordinates (x, y)** show the direction and the distance of the slide. The *x*-coordinate tells how far right or left to slide the figure, and the *y*-coordinate tells how far up or down it moves. The new figure is called an **image**.

Example

Translate the triangle below using the slide coordinates $(3, -2)$.

STEP 1 Add the slide coordinates to the points in the figure.
Add *x*-coordinates together and then add *y*-coordinates to find the new points.
For example, if point *A* is $(1, 3)$, then the new point, A', is $(1 + 3, 3 + -2) = (4, 1)$.
The coordinates $A(1, 3)$, $B(5, 3)$, $C(3, 5)$ become $A'(4, 1)$, $B'(8, 1)$, $C'(6, 3)$.

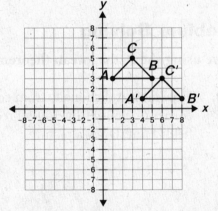

STEP 2 Draw the *image* on the coordinate plane.

ON YOUR OWN

Translate the rectangle *MNOP* by sliding it $(-2, 4)$.

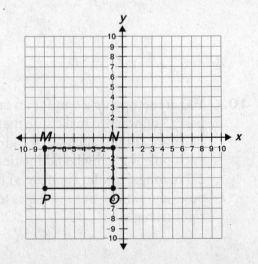

Practice

Building Skills

Solve each translation problem.

Use this diagram for problems 1–4.

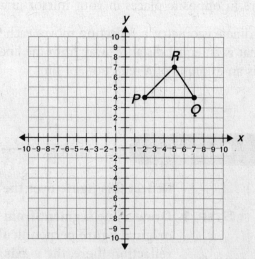

1. What are the points of the new image after a $(-3, -1)$ slide.

2. Draw the figure using the slide coordinates $(2, 0)$.

3. What are the points of the image after a $(2, 6)$ slide?

4. Draw the figure using the slide coordinates $(0, -3)$.

Problem Solving

Use translations to solve these problems.

5. A roller coaster car moves from the ground to the top of a hill 14 m left and 10 m up from the starting place. If the point it begins at is $(0, 0)$, what are the coordinates of the roller coaster's new point?

6. An ice skater slides 10 ft left and 5 ft forward. If the skater started 3 ft left and 8 ft in front of a gate, how far is he from the gate after the slide? Let the gate be at $(0, 0)$.

7. You are riding a canoe from the dock and go 50 ft north and 30 ft east, and then you go 20 ft south and 15 ft east. How far from the dock are you after the two translations (moves)?

8. A map of the city shows 10 blocks across labeled 1H–10H and 10 blocks going up and down labeled 1V–10V. If you start at block (2H, 5V) and walk 3 blocks east and 2 blocks south, where do you end up?

LESSON 32 Reflections

Look at yourself in the mirror. You may notice that your right and left sides are in opposite places in your mirror image.

In coordinate geometry, a **reflection** moves each point in a figure to a place that is an equal distance away from the line of reflection. A mirror image is an example of a reflection, or flip.

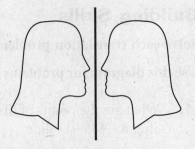

Example

Reflect this figure over the *y*-axis.

STEP 1 Draw a line segment from each point on the original figure perpendicular to the line of reflection (here, the *y*-axis).

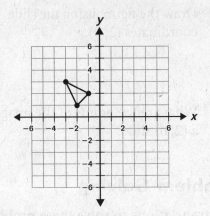

STEP 2 Draw the same line segments on the other side of the reflection line. Connect the points.

STEP 3 Plot the coordinates. Locate the points of the new image on the coordinate plane: (1, 2), (2, 1), (3, 3).

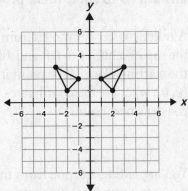

ON YOUR OWN

Reflect this figure over the *y*-axis.

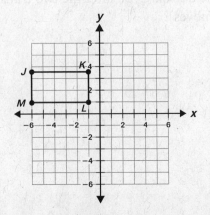

Practice

Building Skills

Solve each reflection problem.

Use this figure to answer questions 1–4.

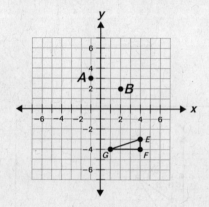

1. If you reflect the point *A* across the *x*-axis, where is its image?

2. If you reflect the point *B* across the *x*-axis, where is its image?

3. If you reflect triangle *EFG* across the *y*-axis, what are the coordinates of the image?

4. If you reflect triangle *EFG* across the *x*-axis, what are the coordinates of the image?

Use this figure to answer questions 5 and 6.

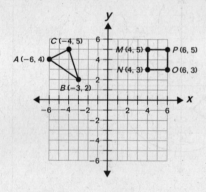

5. If you reflect triangle *ABC* across the *x*-axis, where are the corners of the new image?

6. If you reflect square *MNOP* across the *y*-axis, where are the corners of the new image?

Problem Solving

Solve using reflections.

7. You raise your left hand in front of a mirror. In the image, would you see your right or left hand?

8. A camera flips the image over a line when it records it on film. When the film is developed, the image is flipped back over the same line. Explain why this double flip gives back the original view.

LESSON 33   Rotations

In geometry, a **rotation** is a figure that has been turned around a point. Clock hands are an example of figures being rotated.

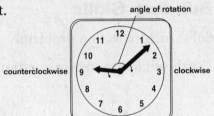

To rotate a figure, you need to know three things:

- the **turn center**, also called the **point of rotation**
- the direction—clockwise or counterclockwise
- the angle of rotation

Example

Rotate triangle *ABC* 180° clockwise around the point of rotation (origin).

STEP 1 Put your pencil point at the origin, (0, 0). Lay your pencil down on the page so that one end is on the origin and the rest of the pencil aligns with the triangle.

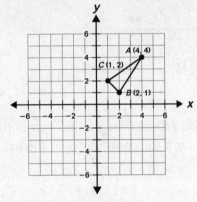

STEP 2 Turning the pencil clockwise, imagine the figure turns 180°. Use the tip of the pencil at the origin as a pivot point and twirl the pencil around.

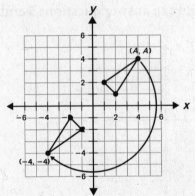

ON YOUR OWN

Rotate the figure 90° counterclockwise.

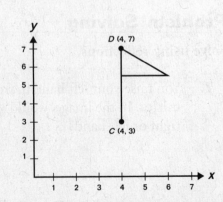

Name _____ Date _____

Practice

Building Skills

Solve each rotation problem.

Use this diagram to answer questions 1–2.
Find the coordinates of each image.

1. Rotate triangle *ABC* around point *C* 90° clockwise.

2. Rotate triangle *DEF* around the origin 180° counterclockwise.

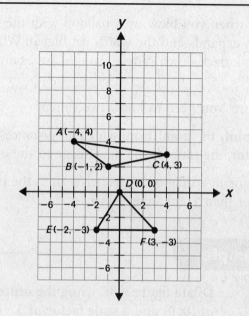

3. Rotate the figure below 90° clockwise.

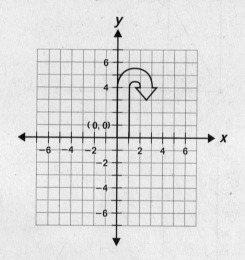

4. Rotate the letter *C* 180° clockwise.

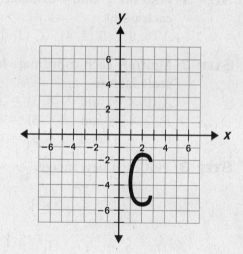

Problem Solving

Solve using rotations.

5. How much of a rotation is the movement of a clock hour hand from 2:00 to 4:00?

6. If a driver needs to make a u-turn, what should be the angle of rotation—90°, 180°, or 360°? Explain.

7. The volume knob on a stereo has 10 volume marks evenly spaced around 270°. From the 0 mark, where would the 11th mark go in this rotation?

8. There are about 12 hours from sunrise to sunset. If the rotation of the sun is about 180° in that time, how many degrees does the sun appear to rotate in one hour?

LESSON 34 Dilations

What happens when you blow up a balloon with the words *Happy Birthday* on it? The balloon expands and the words get bigger. When you let the air out, the words shrink back down. These changes are examples of **dilations,** or stretches.

To dilate a figure, you need to know two things:

- the **center point,** the point from which all distances are measured
- the **scale factor,** the ratio of the new image to the original figure

If the scale factor in a dilation is greater than 1, the image expands. If the scale factor is smaller than 1, the image shrinks.

Example

Dilate figure *ABC* using the center of (0, 0) and a scale factor of 3.

STEP 1 Find the *x*- and *y*-distance(s) to each point.
A (0, 0) B (2, 4) C (2, 0)

STEP 2 Multiply each coordinate by the scale factor.
A (0 × 3) and (0 × 3) = A' (0, 0)
B (2 × 3) and (4 × 3) = B' (6, 12)
C (2 × 3) and (0 × 3) = C' (6, 0)

STEP 3 Plot each new image.

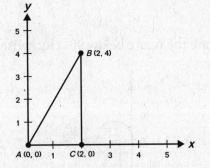

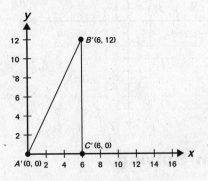

ON YOUR OWN

Dilate figure *ABC* using a scale factor of 2.

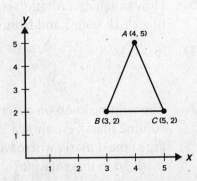

Name _____ Date _____

Practice

Building Skills

Solve each dilation problem. Use this diagram to answer questions 1–4.

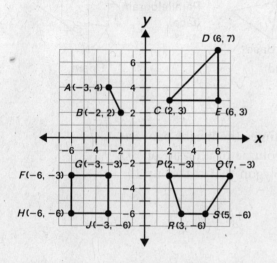

1. Where is the image of triangle *CDE* after a dilation with the center at $(0, 0)$ and a scale factor of 2?

2. Where is the image of square *FGJH* after a dilation with the center at $(0, 0)$ and a scale factor of 0.5?

3. Where is the image of trapezoid *PQSR* after a dilation with the center at $(0, 0)$ and a scale factor of 0.25?

4. Where is the image of segment $\overline{AB}$ after a dilation with the center at $(0, 0)$ and a factor of $\frac{1}{2}$?

Use this diagram to answer questions 5 and 6.

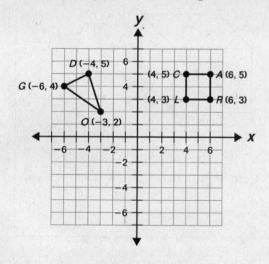

5. Where is the image of triangle *DOG* after a dilation with the center at $(0, 0)$ and a scale factor of 2?

6. Where is the image of rectangle *CARL* after a dilation with the center at $(0, 0)$ and a scale factor of 0.5?

Problem Solving

Use dilations to solve these problems.

7. A word on an empty balloon is 2 in. long. After you blow up the balloon, the word expands by a scale factor of 8. How long is the word on the inflated balloon?

8. Your TV has a picture-in-picture option. The smaller picture is 5 in. wide. The larger picture is a dilation centered in the middle of the small picture with a scale factor of 6. How wide is the large-screen picture?

LESSON 35 Properties of Solids

Solids are three-dimensional shapes. These solids have **faces** that are polygons. **Prisms** have two congruent bases that are parallel polygons. The sides are all parallelograms.

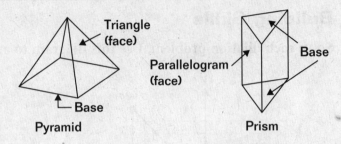

To find the **volume** of a prism, you multiply the area of the base times the height between the bases. Each corner point is called a vertex.

A **pyramid** has only one base and the sides are all triangles.

The volume of a pyramid is equal to $\frac{1}{3}$ of the area of the base times the height between the base and the vertex that is not on the base. So to find the volume of a prism or pyramid, you need to remember how to find the area of the polygon that makes up the base. Here are the general volume formulas:

prism = area of the base × height or V = Bh

pyramid = $\frac{1}{3}$ × area of the base × height or V = $\frac{1}{3}$ Bh

The units of volume are cubed, as in.^{3} or m^3 or ft^3.

Example

Find the volume of this rectangular prism.

STEP 1 Find the area of the base.
The base is a rectangle.
Its area = $l \times w$ = 3 cm × 5 cm = 15 cm^2.

STEP 2 Find the height.
The height is 4 centimeters.

STEP 3 Use the correct formula to find the volume.
The solid is a prism, so Volume = 15 cm^2 × 4 cm = 60 cm^3.

The volume is 60 cm^3.

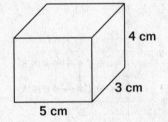

ON YOUR OWN

**Find the volume of this prism
that has right triangle bases.**

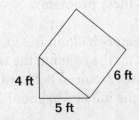

Name _____   Date _____

Practice

Building Skills

Find the volume of each solid.

1.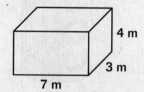

2.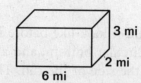

3.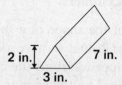

4.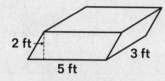

5.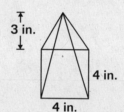

6.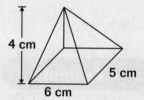

7.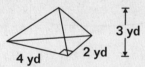

8.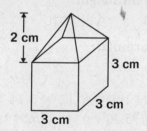

9.

Problem Solving

Solve using properties of solids.

10. You buy your little sister a wading pool. It is a 4-ft by 6-ft rectangular prism. How many cubic feet of water would it take to fill it 1.5 ft deep?

11. Your square cake pans are all 9 in. by 9 in. by 2 in. high. What will be the volume of the cake if you make three layers using these pans (if all pans are filled to the top)?

12. A box for a granola bar is shaped like a triangular prism. The triangles are 2 in. wide and 1.5 in. high, and the triangular bases are 10 in. apart. What is the volume of the box?

13. The Great Pyramid of Giza is a square pyramid with base sides of about 230 m and a height of about 150 m. Using these measurements, what is its volume?

Name _____   Date _____

LESSON 36 Surface Area Using Nets: Cylinder

When you peel an orange or a banana, you take a **three-dimensional shape** and end up with a peel that can be laid out flat. That is the basic idea behind **nets.** A net is the flattened out peel of a 3-D shape. In this lesson you will look at the nets of **cylinders.**

Think of the shape of a soda can. What did it look like before it was put together? It had to have a circular top, a circular bottom, and a rectangle that was wrapped around the circles to create the side of the can. The total **surface area** would be the area of the two circles plus the area of the rectangle.

You know the formula for the area of a circle. It is $A = \pi \times r^2$. You also know the formula for the area of a rectangle. It is $A = l \times w$. By using these two formulas and the idea of nets, you can find the surface area of any cylinder.

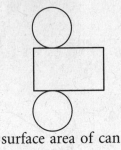

surface area of can

Example

Find the surface area of this cylinder. Use 3.14 for π.

STEP 1 Find the radius and the height.
$r = 4$ cm, $h = 3$ cm

STEP 2 Put them into the formula.

$SA = 2\pi r^2 + 2\pi rh$
$SA = 2 \times (3.14 \times (4 \text{ cm})^2) + (2 \times 3.14 \times 4 \text{ cm}) \times 3 \text{ cm}$
$\quad = 100.48 \text{ cm}^2 + 75.36 \text{ cm}^2 = 175.84 \text{ cm}^2$

The surface area of the cylinder is 175.84 cm^2.

ON YOUR OWN

What is the surface area of a cylinder with radius 2 in. and height 4 in.? Use 3.14 for π.

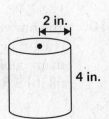

Practice

Building Skills

Find the surface area. Use 3.14 for π.

1.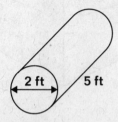
2 ft   5 ft

2.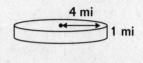
4 mi   1 mi

3.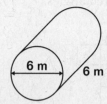
6 m   6 m

4.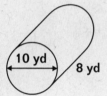
10 yd   8 yd

5.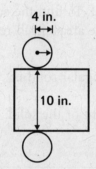
4 in.   10 in.

6.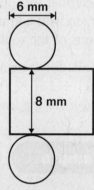
6 mm   8 mm

7.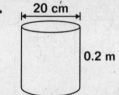
20 cm   0.2 m

8.
4 ft   6.28 ft

9.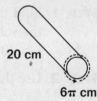
20 cm   6π cm

Problem Solving

Solve using cylinder surface area. Use 3.14 for π.

10. A juice can is about 6 in. high and 2 in. in diameter. How much metal is needed to make a can with these dimensions?

11. You are packing up a cylindrical drum to take to a show. It has a diameter of 18 in. and a height of 6 in. How much bubble wrap will you need to cover it?

12. A muffin is almost cylindrical, with a height of 4 cm and a diameter of 10 cm. How much area is there to cover in one well of a muffin pan with butter spray?

13. You and your father are going to wrap your water heater to insulate it and save money. It is 4 ft high and 2 ft in diameter. How much insulation will it take to cover the sides and the top only?

LESSON 37 Surface Area Using Nets: Rectangular Prisms

This solid is called a rectangular prism, and as you learned with cylinders, nets can be used to find the surface area. If you take a cardboard box and cut it along the edges, you will get a net made up of six rectangles.

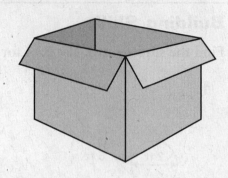

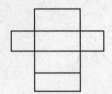

Each of the shapes in the net is a rectangle, and each one is the same as (or congruent to) the rectangle that was opposite it on the solid. To find the surface area of the rectangular prism, you add up all of the rectangle areas. Each rectangle has two of the three dimensions of the solid.

The total formula for the surface area of a rectangular solid is:
$$SA = 2\,(l \times w) + 2\,(l \times h) + 2\,(w \times h)$$
where l, w, and h are the length, width, and height of the solid.

Example

Find the surface area of the box below.

STEP 1 Find the length, width, and height of the solid.
$l = 10$ cm, $w = 3$ cm, and $h = 2$ cm

STEP 2 Use the formula.
$$SA = 2(10 \text{ cm} \times 3 \text{ cm}) + 2(10 \text{ cm} \times 2 \text{ cm}) + 2(3 \text{ cm} \times 2 \text{ cm})$$
$$= 2(30 \text{ cm}^2) + 2(20 \text{ cm}^2) + 2(6 \text{ cm}^2) = 112 \text{ cm}^2$$

The surface area of the box is 112 cm^2.

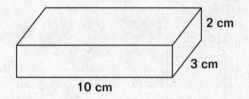

ON YOUR OWN

Find the surface area of the box below.

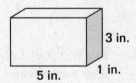

Practice

Building Skills

Find the surface area.

1.
2 m
2 m
2 m

2.
5 yd
2 yd 1 yd

3.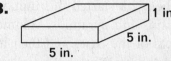
1 in.
5 in.
5 in.

4.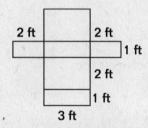
2 ft   2 ft
1 ft
2 ft
1 ft
3 ft

5.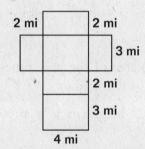
2 mi   2 mi
3 mi
2 mi
3 mi
4 mi

6.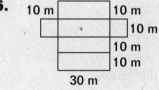
10 m   10 m
10 m
10 m
10 m
30 m

Problem Solving

Use nets to solve these problems.

7. A shipping box with a total area of 350 in.^{2} has a length of 15 in. and a width of 10 in. How high is the box?

8. You are painting a toy box for your nephew. The toy box is a rectangular prism that is 2 ft wide by 3 ft long by 2 ft high. How much surface do you need to paint (leave out the bottom)?

9. How many cm^2 of wrap will it take to cover a CD case that is 1 cm high by 10 cm wide by 12 cm long?

10. You have enough paint to cover 400 sq ft. Will it be enough to paint the ceiling and walls of a room that is 10 ft wide by 10 ft long, by 8 ft high (minus 15 ft^2 for a door and window)? Explain.

11. You are asked to cover your textbook. Your textbook is 10 in. long, 8 in. wide, and 1 in. high. What is the surface area of the front, the spine, and the back cover?

12. You are wrapping a present in a box that is 2 ft by 1 ft by 1 ft. How much paper would it take to cover the box if you do not overlap any of it?

Name _____ Date _____

LESSON 38 Volume of a Cylinder

You want to know how much a can of soup or soda holds. But if the base is not a rectangle or triangle, how can you find out how much the can holds?

When you started this unit, you learned about the volumes of prisms and pyramids. Volumes of those solids depended on the areas of the bases and their heights. Cylinders use the same idea, but the bases are both circles. Once again, because these are circles, you need to use the number π (called pi and equal to about 3.14).

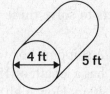

$A = \pi \times r^2$

$C = 2 \times \pi \times r$

The volume of a cylinder is equal to the area of the base times the height between the circular bases.

The formula is: Volume $= \pi \times$ radius$^2 \times$ height or $V = \pi \times r^2 \times h$

Example

What is the volume of the cylinder shown? Use 3.14 for π.

STEP 1 Find the radius and height of the cylinder.
The radius is half the diameter ($r = 2$ ft), and the height is 5 ft.

STEP 2 Use the formula.
$V = 3.14 \times (2\ \text{ft})^2 \times 5\ \text{ft} = 3.14 \times 4\ \text{ft}^2 \times 5\ \text{ft} = 62.8\ \text{ft}^3$

The volume of the cylinder is 62.8 ft^3.

ON YOUR OWN

Find the volume of the cylinder shown.

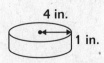

4 in.
1 in.

© Harcourt Achieve Inc. All rights reserved.

84

Lesson 38
Measurement and Geometry, SV 0437-9

Practice

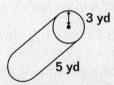
Radius is half the diameter.

Building Skills

Find the volume. Use 3.14 for π.

1.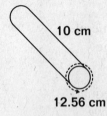
10 cm

12.56 cm

2.
2 ft

6 ft

3.
3 yd

5 yd

4.
10 cm

12 cm

5.
5 cm

31.4 cm

6.
15 mm

18.84 mm

Problem Solving

Use volume to solve these problems. Use 3.14 for π.

7. A can of soup measures 10 cm high by 6 cm in diameter. What is the volume of soup it contains?

8. A bucket of popcorn at the movies is a cylinder with a radius of 6 in. and a height of 10 in. What volume of popcorn can it hold?

9. A subway tunnel is a cylinder with a radius of 4 m and a length between the ends of 1,000 m. What is the volume of dirt workers need to remove to make this tunnel?

10. A can of tennis balls has a 3-in. diameter and is 10 in. high. What is the volume of space inside?

11. Your trash can is 15 in. high and has a diameter of 12 in. What is the volume of trash that you can put inside the trash can?

12. You measure a can that holds 16 oz of juice. It is 14 cm tall and has a diameter of 6 cm. How much space is needed to hold 16 oz?

LESSON 39   Volume of a Cone

Just as a cylinder is like a prism, a **cone** is similar to a pyramid. The
difference in the formula for the volume of a cylinder and a prism is
in their bases. The base of a prism is a polygon. The base of a cylinder
is a circle. A cone is just a pyramid with a circle for a base.

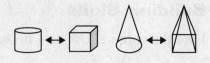

Because a cone is a pyramid with a circle base, the formulas for finding the volume of
a cone and a pyramid are almost the same. Remember, the general formula for the
volume of a pyramid is $V = \frac{1}{3} \times$ area of the base $\times$ height.

This formula is the same for cones, but the area of the base is a circle, and so it equals
π times the radius squared. By changing the pyramid formula slightly, you will have
the formula for cones.

Volume of a cone $= \frac{1}{3} \times \pi \times r^2 \times h$

The height is always the distance between the circle and the top point of the cone, no
matter how the cone is turned.

Example

Find the volume of the cone. Use 3.14 for π.

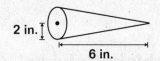

STEP 1 Find the radius and the height.
$r = 2$ in, and $h = 6$ in.

STEP 2 Use the formula.
Volume of a cone $= \frac{1}{3} \times 3.14 \times r^2 \times h = \frac{1}{3} \times 3.14 \times (2 \text{ in.})^2 \times 6 \text{ in.}$
$= \frac{1}{3} \times 3.14 \times 4 \text{ in.}^2 \times 6 \text{ in.} = \frac{1}{3} \times 3.14 \times 24 \text{ in.}^3$
$= 25.12 \text{ in.}^3$

The volume of the cone is 25.12 in.^{3}

ON YOUR OWN

What is the volume of this cone?

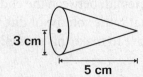

Practice

Building Skills

Find the volume. Use 3.14 for π.

1.
   3 km
   6.28 km

2.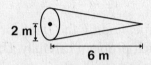
   2 m
   6 m

3.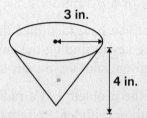
   3 in.
   4 in.

4.
   15 ft
   8 ft

5.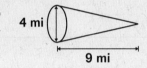
   4 mi
   9 mi

6.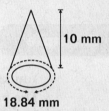
   10 mm
   18.84 mm

Problem Solving

Solve. Use 3.14 for π.

7. How many cubic centimeters of ice cream fit in a cone with a diameter of 4 cm and a height of 12 cm?

8. An insect called a sand lion digs holes that are cone-shaped in the sand. How much sand did the sand lion need to clear out to make a hole that is 20 mm deep and has a radius of 12 mm?

9. What is the volume of a traffic cone that is 3 ft high and has a diameter of 1 ft?

10. What is the volume of a funnel that has a 4 in. diameter and is 3 in. high?

11. At the costume party, Sandra wore a pointed hat. She needed something to keep her wallet in, so she turned over her hat and used the hat as a purse. If the circumference of the hat is about 12.56 in., and it is 24 in. high, what is the volume of the hat?

12. A cone-shaped pile of sand is 5 ft high, and has a diameter of 6 ft. How many cubic feet of sand are in this pile?

LESSON 40 Solving Two-Step Volume Problems

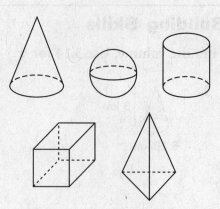

When solving a volume problem, try to imagine the shapes. Does the shape have all flat sides and right angles like a cube or a box? Or are there curves like a cylinder or sphere? Is there a point like a cone or a pyramid?

Once you have identified the correct shape, finding the right formula is the key to finding the answer. Read each problem carefully. Look for given values. Find any key words that may help you.

• If the problem has a radius, it must be a cylinder, cone or sphere.
• Cubes, cylinders, pyramids, wedges, and cones have a base and a height.

When you know the shape and the formula, you are ready to do the two-step solution.

Example

You know that you will need 200 in.^{3} of paint to cover the walls in a room. The paint is sold in cans that are 8 inches tall and have a radius of 3 inches. How much paint is in these cans? Is one can enough to paint this room?

STEP 1 Think of the shape in the problem.
Paint cans have a circular base and straight sides. The problem also lists both a radius and a height. This must be a cylinder.

STEP 2 Find the correct volume formula for this shape.
The volume of a cylinder is $V = \pi r^2 h$.

STEP 3 Solve the area part of the formula first.
The area of the circle or base is πr^2.
$3.14 \times 3^2 = 28.26$ square inches

STEP 4 Multiply the area times the height.
$V = 28.26$
$V = 28.26 \times 8$ in.

STEP 5 Solve the equation.
$V = 3.14 \times 3^2 \times 8 = 3.14 \times 9 \times 8 = 226.08$ in.^{3}

226.08 in.^{3} > 200 in.^{3}, so you will have enough paint.

ON YOUR OWN

Orange juice is sold in a paper carton with a square base that is 2 in. on each side. If the carton's volume is 24 in.^{3}, what is the height of this container?

Practice

Building Skills

The volume of a rectangular pyramid is $V = \frac{1}{3} Bh$. B = area of base = $l \times w$. Fill in the table below.

Base	Height	Volume
1. 3 in. $\times$ 4 in.	6 in.	
2. 7.5 cm $\times$ 5 cm	10 cm	
3. $\frac{1}{2}$ ft $\times$ 2 ft	3 ft	
4. 20 in. $\times$ 7 in.	12 in.	

The volume of a cone is $V = \frac{1}{3} \pi r^2 h$. Fill in the table below. Use 3.14 for π.

Radius	Height	Volume
5. 8 cm	16 cm	
6. 2 m	3 m	
7. 0.5 ft	4 ft	
8. 1 yd	6 yd	

Problem Solving

Write an equation and solve. Use 3.14 for π.

9. You need 300 in.^{3} of cleaning solution to wash the outside of a building. The solution is sold in cans of different heights. Each can has a radius of 3 in. How tall should the can be to hold the amount of solution that you need?

10. You buy a box of paper clips along with other school supplies. This box is 3 in. long and 2 in. wide. If the total volume is 3 in.^{3}, how tall is this box?

11. Dr. Chang buys a fish tank for his office. The tank is 5 ft long, 4 ft wide, and 3 ft tall. Is the tank large enough for fish that require 50 ft^3 of water?

12. After lunch, you buy some frozen yogurt in a waffle cone. The radius of this cone is 1.5 in. and is 6 in. deep. If it is filled just to the top, how much yogurt can this waffle cone hold?

LESSON 41 Two-Dimensional Views of Three-Dimensional Objects

Have you ever noticed that an object can look very different depending upon how you look at it? Consider the views of a cone. The cone appears to be different shapes depending upon how you view it.

cone

If you look at a cone from the side, it looks like a triangle.

view from side

If you look down on a cone from above, it looks like a circle.

view from above

What do you think a cone would look like if you viewed it from the bottom?

There are ways to recognize two-dimensional views of three-dimensional objects.

Example

Look at how the cubes are stacked in the solid. Which pattern shows you the view of the solid as you look down from above?

(a)   (b)   (c)

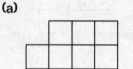 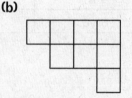

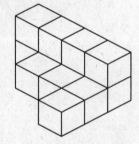

STEP 1 Count off the top surface of each cube that you can see.
If you are looking for a side view, you count off the surface of each cube facing that side.
If you are looking for the front view, you count off the surface of each cube facing you.

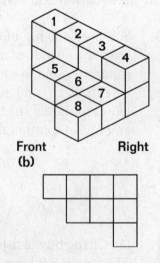
Front   Right
(b)

STEP 2 Look for the view that has the same number of cubes as in STEP 1.

STEP 3 Make sure that the view has the same pattern of cubes. Think of the stacked cubes as having a base plan. What would the layout look like?

The view looking down from above is choice (b).

(ON YOUR OWN)

What is the left side view of the solid?

Name _____   Date _____

Practice

Building Skills

Use this solid to answer questions 1–4.

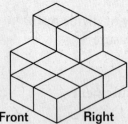

Front   Right

1. Which view shows the object from the left side?

2. Which view shows the object from the front?

3. Which view shows the object from above?

4. Which view shows the object from the right side?

A.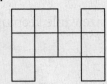
   B.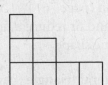
   C.
   D.

Use this solid to answer questions 5–7.

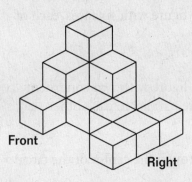

Front

Right

5. Which view shows the object from the left side?

6. Which view shows the object from the front?

7. Which view shows the object from above?

A.
   B.
   C.
   D.

Use this solid to answer questions 8–10.

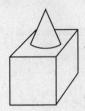

8. Which view shows the object from the left side?

9. Which view shows the object from the front?

10. Which view shows the object from above?

A.
   B.
   C.
   D.

Glossary

acute angle (page 48)
an angle whose measure is less than 90°

acute triangle (page 52)
a triangle in which all angles are acute angles

angle (page 48)
when two lines intersect at a point

area (page 26)
the measure of the space inside a flat figure: area of a parallelogram = $l \times w$; area of a triangle = $1/2\, bh$

base (page 28)
any side in a parallelogram or triangle; the bottom of a three-dimensional shape

capacity (page 32)
the measure of how much of something a container holds

center point (page 76)
the point from which all distances are measured when a figure is dilated

chord (page 56)
a line segment whose endpoints lie on the circle

circle (page 56)
a closed set of infinite points on a line

circumference (pages 56, 58)
the distance around the edge of a circle

complementary angles (page 50)
two angles whose sum is 90°

complex figure (page 30)
the combination of two or more simple figures into one object

cone (page 86)
a three-dimensional figure that has a circular base, a vertex, and a curved surface

congruent (pages 50, 64)
equal in size

congruent figures (page 64)
figures that have the same shape and size

conversion factor (pages 10, 18)
the number that you multiply or divide by to change to another unit of measure

convert (page 10)
change, as with units of measurement

coordinate plane (page 70)
a horizontal number line (the x-axis) crossed with a vertical number line (the y-axis)

corresponding angles (page 50)
two equal angles formed by a line intersecting two parallel lines

cube (page 90)
a three-dimensional figure with six faces, each of which is a square

cylinder (page 80)
a three-dimensional figure with two congruent circular bases that are parallel to each other

diameter (page 56)
a chord that goes through the center of the circle

dilation (page 76)
to expand or reduce a figure in size while keeping the same shape

equilateral triangle (page 52)
a triangle in which all sides are congruent lengths

face (page 78)
a polygon that forms one of the surfaces of a three-dimensional figure

height (page 28)
the distance from the bottom to the top of something

image (page 70)
new figure after a transformation

intersect (page 46)
when lines or planes cross each other

isosceles triangle (page 52)
a triangle in which two sides are congruent lengths

kite (page 54)
a quadrilateral with two equal adjoining sides

line (page 44)
a straight path that extends without end in opposite directions

line of symmetry (page 66)
the line across a figure that shows the symmetry of the figure

line segment (page 44)
a part of a line with two endpoints

mass (page 40)
a measure of the amount of matter in an object

net (page 80)
a two-dimensional pattern that can be folded to form a three-dimensional figure

obtuse angle (page 48)
an angle whose measure is greater than 90°

obtuse triangle (page 52)
a triangle with one obtuse angle

origin (page 70)
the point where the x-axis and y-axis cross

parallel lines (page 46)
lines in the same plane that never intersect

parallelogram (pages 28, 54)
a quadrilateral with two pairs of parallel sides

perimeter (page 24)
the measure of the distance around the outside of a figure

perpendicular lines (page 46)
lines that intersect to form right angles

pi (π) (page 58)
the ratio of the circumference of a circle to its diameter; equal to approximately 3.14

plane (page 44)
a flat surface that goes on forever

point (page 44)
an exact location

point of rotation (page 74)
another term for the turn center

polygon (page 52)
a closed figure formed by three or more line segments that do not cross; a flat closed figure with at least three sides

prism (page 78)
a three-dimensional figure that has two congruent bases that are parallel polygons

protractor (page 48)
an instrument used to measure angles

pyramid (page 78)
a three-dimensional figure in which one face, the base, is a polygon, and the other faces are triangles that meet at a common point, called the vertex

Pythagorean theorem (page 68)
a formula used to find the lengths of sides in a right triangle: $a^2 + b^2 = c^2$

quadrilateral (page 54)
a polygon with four sides, four angles, and four vertices

radius (page 56)
the distance from the center of the circle to the edge

rate (page 22)
a ratio that compares two quantities measured in different units

ray (page 44)
a part of a line that has one endpoint and goes without stopping in one direction only

rectangle (pages 26, 54)
a quadrilateral with four right angles

reflection (page 72)
flipping a figure over a given line

rhombus (page 54)
a quadrilateral with four equal, parallel sides

right angle (page 48)
an angle whose measure is 90°

right triangle (page 52)
a triangle with a right angle

rotation (page 74)
a transformation in which a figure is turned about a point a certain number of degrees

scale (page 20)
a ratio of any length in a scale drawing to the corresponding actual length

scale factor (page 76)
the ratio of a dilated image to the original figure

scalene triangle (page 52)
a triangle in which all the sides are different lengths

similar figures (page 64)
figures that have the same proportions

slide coordinates (page 70)
coordinates that show the direction and the distance of a slide

solid (page 78)
another term for a three-dimensional shape

square (page 54)
a quadrilateral with four equal sides and four right angles

straight angle (page 48)
an angle whose measure is 180°

supplementary angles (page 50)
two angles whose sum is 180°

surface area (page 80)
the sum of the areas of all the faces and base of a prism, cylinder, pyramid, or cone

symmetry (page 66)
a quality of a figure that can be folded or moved so that two parts of the figure match

three-dimensional shape (page 80)
a shape that has height, length, and width; also called a 3-D shape

translation (page 70)
sliding a figure to a new location

trapezoid (page 54)
a quadrilateral with only one pair of parallel sides

triangle (pages 28, 52)
a polygon with three sides, three angles, and three vertices

turn center (page 74)
point around which a figure is rotated

variable (page 22)
a quantity that can change; a letter that stands for a number

vertex (page 48)
the point where lines come together to form an angle; a corner point

vertical angles (page 50)
opposite angles formed by intersecting lines

volume (page 78)
a measure of the amount of space inside an object

weight (page 38)
a measure of how heavy an object is

ASSESSMENT

PAGES 4–8

1. 72 in. ÷ 36 in. = 2 yd
2. 2,670 m ÷ 1,000 m = 2.67 km
3. 508 cm ÷ 2.54 cm = 200 in.
4. 5 km ÷ 1.6 km = 3.125 mi
5. 60 in. + 2 yd = 60 in. + 72 in. = 132 in. = 11 ft
6. 1.8 km − 258 m = 1,800 m − 258 m = 1,542 m
7. $\frac{1}{4}$ mi = 0.25 mi × 5,280 ft = 1,320 ft
8. 1,672 cm ÷ 100 = 16.72 m
9. 120 in. + 2 ft = 10 ft + 2 ft = 12 ft
10. 3.2 km − 145 m = 3,200 m − 145 m = 3,055 m
11. 5 in. × 2.54 = 12.7 cm
12. 1.6 km ÷ 1.6 = 1 mi 13. 16 m × 4 m = 64 m^2
14. 10 in. × 25 = 250 in. 15. 12 ft × 2 ft = 24 ft^2
16. 6 in. × 2.5 ft = 15 ft 17. 305 mi ÷ 5 h = 61 mph
18. 57 mph × $3\frac{1}{2}$ h = 57 × 3.5 = 199.5 mi
19. 500 mi ÷ 133.87 mph = 3.73 h
20. 105 mi × 11 d = 1,155 mi
21. perimeter = 4 m + 3 m + 5 m = 12 m;
    area = $\frac{1}{2}$(4 m × 3 m) = 6 m^2
22. perimeter = 3 ft × 4 = 12 ft; area = 3 ft × 3 ft = 9 ft^2
23. perimeter = 3 yd + 3 yd + 2.2 yd + 2.2 yd = 10.4 yd;
    area = 2.2 yd × 3 yd = 6.6 yd^2
24. perimeter = 10 km + 17 km + 19.72 km = 46.72 km;
    area = $\frac{1}{2}$ (17 km × 10 km) = 85 km^2
25. 10 pt ÷ 2 = 5 qt
26. 200 mL ÷ 1,000 = 0.2 L
27. 2 gal × 3.8 = 7.6 L
28. 7 pt ÷ 2 = 3.5 qt
29. 2,000 mL ÷ 1,000 = 2 L
30. 60 mL ÷ 30 = 2 fl oz
31. 12 lb × 16 = 192 oz
32. 10 kg × 1,000 = 10,000 g
33. 7.5 kg × 2.2 = 16.5 lb
34. 6 tons × 2,000 = 12,000 lb
35. 1,050 g ÷ 1,000 = 1.05 kg
36. 100 g × 0.035 = 3.5 oz
37. parallel because the angles are equal where the lines are intersected by a transversal
38. intersecting but not perpendicular because the angle between them is not 90°
39. intersecting but not perpendicular because the angles between them are not 90°
40. perpendicular because the angle between them is 90°
41. acute because the angle is less than 90°
42. right because the angle is 90°
43. obtuse because the angle is greater than 90°
44. straight because the angle is 180°
45. 180° − 45° = 135°
46. 180° − 72° = 108°
47. $\overline{MO}$, $\overline{RO}$, $\overline{QO}$
48. C = 2 π r = 2 × 3.14 × 4 cm = 25.12 cm
49. r = d/2 = 4 yd/2 = 2 yd
    A = πr^2 = 3.14 × (2 yd)^2 = 3.14 × 4 yd^2 = 12.56 yd^2
50. A = $\frac{1}{2}$(b × h) = $\frac{1}{2}$ (3 ft × 5 ft) = $\frac{1}{2}$ × 15 ft^2 = 7.5 ft^2
51. 5 mi/x = 6 mi/3 mi
    6x = 5 × 3 = 15 mi; x = 2.5 mi
52. angle C
53. (2 − 3, 4 + 4) = (−1, 8)
54. The x-coordinate stays the same. The y-coordinate is the negative of its original value. (5, −3)
55. 90° is one-fourth the way around the graph; (−2, −1).
56. (−1) − 0 = −1; (−4) − 0 = −4
    (−1) × 2 = −2; (−4) × 2 = −8
    0 + (−2) = −2; 0 + (−8) = −8
    (−2, −8)
57. V = b × h/3 = (6 cm × 6 cm) × 9 cm/3 = 324 cm^3/3 = 108 cm^3
58. V = b × h/3 = (πr^2) × $\frac{h}{3}$ = 3.14 × (3 in.)^2 × 4 in./3 = 3.14 × 9 in. × 4 in./3 = 113.04 in.^{3}/3 = 37.68 in.^{3}
59. SA = 2(4 mi × 10 mi) + 2(6 mi × 10 mi) + 2(4 mi × 6 mi) = 80 mi^2 + 120 mi^2 + 48 mi^2 = 248 mi^2
60. SA = 2(π × 10^2) + (2 × π × 10) × 15 m = 1,570 m^2
61. V = l × w × h = 7 ft × 3 ft × 4 ft = 84 ft^3
62. C

LESSON 1

ON YOUR OWN (page 10): 1.5

PAGE 11

1. 150 in. ÷ 12 in. = 12.5 ft 2. 4,400 yd
3. 100 yd × 3 = 300 ft 4. 0.5 mi
5. 880 yd × 3 ft = 2,640 ft ÷ 5,280 ft = 0.5 mi
6. 316,800 in. 7. 600 in. ÷ 36 in. = 16.67 yd
8. 300 in. 9. 3.9 mi × 5,280 ft = 20,592 ft
10. 115.2 in. 11. 264 in. ÷ 12 in. = 22 ft
12. 107.04 in. 13. 46 ft ÷ 3 ft = 15.33 yd
14. 4,400 yd 15. 0.75 ft × 12 in. = 9 in.
16. 14 ft; no, it will not fit under the 11-ft overpass

LESSON 2

ON YOUR OWN (page 12): 2,700 m

PAGE 13

1. 22 cm ÷ 100 cm = 0.22 m
2. 2,500 m
3. 100 mm ÷ 1,000 mm = 0.1 m
4. 520 cm
5. 0.4 km × 1,000 m = 400 m
6. 729 cm
7. 0.6 m × 1,000 mm = 600 mm
8. 0.3475 km
9. 179,000,000 mm ÷ 1,000 mm = 179,000 m ÷ 1,000 m
   = 179 km
10. 760,000 cm
11. 50 m ÷ 1,000 m = 0.05 km
12. 1.28 km
13. 116 cm ÷ 100 cm = 1.16 m
14. 4.877 m
15. 3.389 km × 1,000 m = 3,389 m
16. 182 cm to 216 cm

LESSON 3

ON YOUR OWN (page 14): 2,080 m

PAGE 15

1. 45.7 cm − 320 mm = 457 mm − 320 mm = 137 mm
2. 345 ft
3. 1,370 cm + 42.7 m = 1,370 cm + 4270 cm = 5,640 cm
4. 2,563 yd
5. 0.8 mi + 4,130 ft = 4,224 ft + 4,130 ft = 8,354 ft
6. 2,587 mm
7. 432 in. + 37.25 ft = 432 in. + 447 in. = 879 in.
8. 900 cm
9. 16 yd − 31.5 ft = 48 ft − 31.5 ft = 16.5 ft
10. 1,987 m
11. 30 yd + 15 ft = 90 ft + 15 ft = 105 ft
12. 11,373 m
13. 100 yd + 30 ft + 30 ft = 300 ft + 30 ft + 30 ft = 360 ft
14. 23.5 ft
15. 2.4 mi − 500 ft = 12,672 ft − 500 ft = 12,172 ft
16. 61,776 ft

LESSON 4

ON YOUR OWN (page 16): 7,800 cm

PAGE 17

1. 3,760 mm ÷ 40 = 94 mm
2. 756 ft^2
3. 350 cm × 5.6 m = 3.5 m × 5.6 m = 19.6 m^2
4. 99,000 ft^2
5. 0.5 mi × 28 ft = 2,640 ft × 28 ft = 73,920 ft^2
6. 75 m
7. 1.5 mi ÷ 200 ft = 7,920 ft ÷ 200 ft = 39.6 ft
8. 4 cm
9. 963 yd ÷ 120 in. = 34,668 in. ÷ 120 in. = 288.9 in.
10. 15,768 mm^2
11. 13 ft × 20 ft = 260 ft^2
12. 20 parts
13. 4.5 km × 5 = 22.5 km
14. 150 m
15. 14 ft ÷ 21 = 168 in. ÷ 21 = 8 in.
16. 44 plots

LESSON 5

ON YOUR OWN (page 18): 1,640 ft

PAGE 19

1. 4.5 mi × 1.6 = 7.2 km
2. 13.1 m
3. 50.8 cm × 0.394 = 20 in.
4. 1.04 km
5. 2.4 yd × 3 = 7.2 ft × 0.305 = 2.2 m
6. 4,647 yd
7. 125 ft × 0.305 = 38.1 m ÷ 1,000 = 0.0381 km
8. 0.0788 in.
9. 483 mm ÷ 1,000 = 0.483 m × 3.28 = 1.58 ft = 0.53 yd
10. 0.022 mi
11. 5 km × 0.6 = 3 mi
12. 475.8 mi
13. 16.728 ft
14. 20.29 mpg
15. 3.8 cm
16. 4.575 m, 0.864 m

LESSON 6

ON YOUR OWN (page 20): 17 cm

PAGE 21

1. 15 cm × (40 cm/ 1 cm) = 600 cm or 6 m
2. 100 cm
3. (5 in. /1 in.) × (1.5 mi/1 in.) = 7.5 mi
4. 15 in.
5. (40 ft /1) × (1/20) = 40/20 ft = 2 ft
6. 2.5 ft
7. 40 km = (40,000 m/ 1cm) × (1 cm/ 1 m) = 40,000 cm
8. 10 in.
9. 12 mi ÷ 0.25 mi/1 in. = 48 in.
10. 8.25 mi
11. $\frac{2 \text{ m}}{1} \times \frac{2 \text{ cm}}{1 \text{ m}} = 4$ cm
12. 0.8 ft or 9.6 inches
13. $(\frac{6 \text{ cm}}{1 \text{ cm}}) \times (\frac{1 \text{ km}}{1 \text{ cm}}) = 6$ km
14. 2.9 feet × 12 = 34.8 in.

LESSON 7

ON YOUR OWN (page 22): 192.5 mi

PAGE 23

1. $d = (45)(1.5) = 67.5$ mi
2. 120 m
3. $d = (24)(7) = 168$ ft
4. 900 km
5. $r = 2.5/34 = 0.07$ km/min
6. about 11.7 mi/h
7. $t = 75/60 = 1.25$ h
8. about 18.2 s
9. $d = (25)(15) = 375$ m
10. .13 km/min or 7.8 km/h
11. $d = 250,000$ mi ÷ 3 days = about 83,000 mi
12. about 0.6 h, or about 34 minutes

LESSON 8

ON YOUR OWN (page 24): 120 m

PAGE 25

1. 2 cm + 1 cm + 3 cm + 1 cm = 7 cm
2. 9 ft
3. 5 in. + 12 in. + 13 in. = 30 in.
4. 66 ft
5. 3 cm + 3 cm + 2 cm + 2 cm = 10 cm
6. 42 m
7. 2.4 yd + 3.8 yd + 2.4 yd = 8.6 yd
8. 12 cm
9. 2 in. + 1.5 in. + 1.5 in. + 2 in. + 1 in. = 8 in.
10. 80 in.
11. 90 ft + 90 ft + 90 ft + 90 ft = 360 ft
12. 920 m
13. 7 ft + 7 ft + 4.5 ft + 4.5 ft = 23 ft

LESSON 9

ON YOUR OWN (page 26): 10 cm^2

PAGE 27

1. 1 in. × 2 in. = 2 in.^{2}
2. 3 in.^{2}
3. 2 cm × 4 cm = 8 cm^2
4. 8.25 cm^2
5. 0.5 in. × 0.25 in. = 0.125 in.^{2}
6. 6.5 cm^2
7. 2 ft × 2 ft = 4 ft^2
8. 9 m^2
9. 8.5 ft × 1.5 ft = 12.75 ft^2
10. 144 mi^2
11. 27.3 km × 2 km = 54.6 km^2
12. 80 in.^{2}
13. 120 yd × 53.33 yd = 6,400 yd^2
14. 1.25 mi^2
15. 7 ft × 15.5 ft = 108.5 ft^2, no, the area is greater than 100 ft^2.

LESSON 10

ON YOUR OWN (page 28): 3.75 cm^2

PAGE 29

1. $\frac{1}{2}$(2.6 cm × 0.9 cm) = 1.17 cm^2
2. 9 in.^{2}
3. 5 cm × 1 cm = 5 cm^2
4. 12.5 in.^{2}
5. 5 m × 4 m = 20 m^2
6. 4 ft^2
7. (4.8 yd × 3.6 yd)/2 = 8.64 yd^2
8. 60 in.^{2}
9. (20 m × 9.6 m)/2 = 96 m^2
10. 24 m^2
11. (20 ft)(30 ft) = 600 ft^2
12. 4.2 yd^2
13. The rectangle has the greater area.
    $A_{triangle} = \frac{1}{2}(3)(20) = 30$ in.^{2}
    $A_{rectangle} = (4)(10) = 40$ in.^{2}

LESSON 11

ON YOUR OWN (page 30): perimeter = 19 ft
area = 16.5 ft^2

PAGE 31

1. 9.6 yd
2. 64 in.
3. 27.2 m
4. Perimeter = 1.5 in. + 1.1 in. + 2 in. + 1 in. = 5.6 in.
   Area = (1.5 in. × 1 in.) + ($\frac{1}{2}$ × 0.5 in. × 1 in.) = 1.75 in.^{2}
5. Perimeter = 82 cm; Area = 375 cm^2
6. Perimeter = 6 ft + 6 ft + 5 ft + 5 ft + 6 ft = 28 ft
   Area = (6 ft × 6 ft) + $\frac{(6\,ft \times 4\,ft)}{2}$ = 36 ft^2 + 12 ft^2 = 48 ft^2
7. Perimeter = 58 in.; Area = 216.75 in.^{2}
8. Perimeter = 80 m; Area = 250 m^2
9. Perimeter = 86 ft; Area = 120 ft^2
10. Area = (30 ft)(3 ft) + (10 ft)(3 ft) = 120 ft^2;
    Perimeter = 10 ft + 10 ft + 10 ft + 10 ft + 10 ft + 3 ft + 3 ft
    + 3 ft + 13.5 ft + 13.5 ft = 86 ft

LESSON 12

ON YOUR OWN (page 32): 3c

PAGE 33

1. 5 pt ÷ 2 pt/qt = 2.5 qt
2. 12 qt
3. 16 pt ÷ 2 pt/qt 4 qt/gal = 2 gal
4. 10 c
5. 20 fl oz ÷ 8 fl oz/c = 2.5 c
6. 16 fl oz
7. 28 qt ÷ 4 qt/gal = 7 gal
8. 1.5 pt
9. 1.5 gal × 4 qt/gal × 2 pt/qt × 2 c/pt = 24 c
10. 1.25 gal
11. 5 gal × 4 qt/gal = 20 qt
12. 24 loads
13. 70 fl oz ÷ 8 fl oz/c = 8.75 c
14. 128 c
15. 14 quarts; 14 qt ÷ 4 qt/gal = 3.5 gal
16. more water; 16 cups in 1 gal

LESSON 13

ON YOUR OWN (page 34): 0.45 L

PAGE 35

1. 2,000 mL ÷ 1,000 = 2 L
2. 4,700,000 mL
3. 5,300 mL ÷ 1,000 = 5.3 L
4. 0.4 L
5. 537 L × 1,000 = 537,000 mL
6. 49,400 mL
7. 23,000 mL ÷ 1,000 = 23 L
8. 57.89 L
9. 12 L × 1,000 = 12,000 mL
10. 20 L
11. 2 L × 1,000 = 2,000 mL
12. 0.5 L
13. 355 mL × 6 = 2,130 mL ÷ 1,000 = 2.13 L
14. 0.325 L
15. 0.8 L × 1,000 = 800 mL
16. 0.5 L

LESSON 14

ON YOUR OWN (page 36): 91.2 L

PAGE 37

1. 15 mL × 0.204 tsp./1 mL = 3.06 tsp.
2. 0.08 gal
3. 1.5 c × 8 fl oz/c ÷ 33.8 fl oz = 0.36 L
4. 1,598 mL
5. 18.7 L × 0.26 gal/L = 4.86 gal
6. 3,891,200 mL
7. 16 tsp. × 4.9 mL/tsp. = 78.4 mL
8. 2.12 L
9. $\frac{8}{9}$ L = 1.88 pt
10. 0.68 L
11. 12 gal × 3.8 L = 45.6 L
12. 0.34 c or about 1/3 cup
13. 8,000 L ÷ 3.8 L = 2,105.26 gal
14. 2 L bottle; 2 L > 1.9 L
15. 2.5 qt ÷ 4 qt = 0.625 gal × 3.8 L = 2.375 L ÷ 1,000 mL
    = 2,375 mL
16. It is less than 0.5 L.

LESSON 15

ON YOUR OWN (page 38): 1.4 tons

PAGE 39

1. 3 tons $\times$ 2,000 = 6,000 lb
2. 4 lb
3. 2 lb $\times$ 16 = 32 oz
4. 4.5 tons
5. 10 lb $\times$ 16 = 160 oz
6. 8 lb
7. 24 oz $\div$ 16 = 1.5 lb
8. 1.5 tons
9. 4 tons $\times$ 2,000 = 8,000 lb
10. 16,000 oz
11. 5 lb $\times$ 16 = 80 oz
12. 4 oz
13. 1.2 tons $\times$ 2,000 = 2,400 lb
14. 2.5 tons
15. They have an equal price per ounce; $2.40 $\div$ 12 oz = $0.20/oz; $3.20 $\div$ 16 oz = $0.20/oz
16. 13,000 lb

LESSON 16

ON YOUR OWN (page 40): 24,000 mg

PAGE 41

1. 1.5 kilograms $\times$ 1,000 grams = 1,500 grams
2. 4.5 kg
3. 1,250 mg $\div$ 1,000 = 1.25 g
4. 2,500 g
5. 6 g $\times$ 1,000 = 6,000 mg
6. 3,000 g
7. 500 mg $\div$ 1,000 = 0.5 g
8. 8 kg
9. 1.4 kg $\times$ 1,000 = 1,400 g
10. 500 mg
11. 2 kg $\times$ 1,000 = 2,000 g
12. 0.527 kg
13. 60 mg $\div$ 1,000 = 0.06 g
14. 680 g
15. 750 g $\times$ 4 $\div$ 1,000 = 3 kg
16. 0.25 g

LESSON 17

ON YOUR OWN (page 42): 909 kg

PAGE 43

1. 7 oz $\div$ 0.035 oz = 200 g
2. 2,000 g
3. 11 lb $\div$ 2.2 = 5 kg
4. 2,857 g
5. 2 tons $\times$ 2,000 = 4,000 lb $\div$ 2.2 = 1,818.2 kg
6. 35.2 lb
7. 6.6 lb $\div$ 2.2 = 3 kg $\div$ 1,000 = 3,000 g
8. 0.99 lb
9. 100 g $\times$ 0.035 oz = 3.5 oz
10. 0.91 kg
11. 60 g $\times$ 0.035 oz = 2.1 oz
12. 11 lb
13. 5,280 lb $\div$ 2.2 = 2,400 kg

14. 114.3 g
15. 50 lb $\div$ 2.2 = 22.7 kg
16. No. 6,000 kg $\times$ 2.2 kg/lb = 13,200 lb

LESSON 18

ON YOUR OWN (page 44): Answers will vary.

PAGE 45

1. 5: $\overline{BC}$, $\overline{BD}$, $\overline{CD}$, $\overline{AC}$, $\overline{AD}$
2. 4: A, B, C, D
3. 5 line segments
4. 5 points
5. 17 rays: AD, BD, ED, DE, BC, CB, CA, EA, CE, AE and the rays pointing outward from points A, B, C, D (2), E (2)
6. 3 lines: AE, BC, DE
7. 8 points at the corners
8. 8 line segments make up the sides
9. $\overrightarrow{PQ}$, $\overrightarrow{RQ}$, and $\overrightarrow{PR}$
10. four of $\overrightarrow{PQ}$, $\overrightarrow{RQ}$, $\overrightarrow{RP}$, $\overrightarrow{PR}$, $\overrightarrow{QR}$, $\overrightarrow{QP}$ and the rays going from P to the left and R to the right
11. $\overleftrightarrow{PQ}$, $\overleftrightarrow{QP}$, $\overleftrightarrow{RQ}$, $\overleftrightarrow{QR}$, $\overleftrightarrow{RP}$, $\overleftrightarrow{PR}$
12. $\overrightarrow{RP}$
13. Answers might include line segments $\overline{BC}$, $\overline{AC}$, $\overline{CD}$, $\overline{ED}$, $\overline{AE}$, $\overline{AB}$, $\overline{BE}$; rays $\overrightarrow{CA}$, $\overrightarrow{ED}$; points A, B, C, D, E. There are no lines.

LESSON 19

ON YOUR OWN (page 46): Intersecting but not perpendicular.

PAGE 47

1. Intersecting but not perpendicular.
2. Perpendicular; the lines intersect and form 4 corners.
3. Parallel; the lines are in the same plane and do not intersect.
4. Intersecting but not perpendicular.
5. Parallel; the lines are in the same plane and do not intersect.
6. Intersecting but not perpendicular.
7. Intersecting but not perpendicular; the lines intersect if extended but they do not form 4 corners.
8. Perpendicular; the lines intersect and form 4 corners.
9. Intersecting but not perpendicular.
10. Perpendicular; the lines intersect and form 4 corners.
11. Parallel.
12. Perpendicular.
13. These lines are not in the same plane, so they do not intersect and are not parallel.
14. Perpendicular.
15. Parallel.
16. Parallel.

LESSON 20

ON YOUR OWN (page 48): obtuse

PAGE 49

1. Obtuse; the measure of the angle is between 90° and 180°.
2. Right; the measure of the angle is 90°.
3. Acute; the measure of the angle is less than 90°.
4. Obtuse; the measure of the angle is between 90° and 180°.
5. Right; the measure of the angle is 90°.
6. Acute; the measure of the angle is less than 90°.
7. Obtuse; the measure of the angle is between 90° and 180°.
8. Straight; the measure of the angle is 180°.
9. Obtuse; the measure of the angle is between 90° and 180°.
10. 48°; the measure of the angle is less than 90°.
11. 132°; the measure of the angle is between 90° and 180°.
12. Straight; the measure of the angle is 180°.
13. $\frac{360°}{30°} = 12$ pieces   14. $180° - 45° = 135°$

LESSON 21

ON YOUR OWN (page 50): 54°

PAGE 51

1. 42°, ∠1 and ∠3 are vertical angles; ∠3 and ∠7 are corresponding angles; ∠1 = ∠3 = ∠7 = 42°
2. ∠5 and ∠8 are supplementary, so ∠8 = 180° − 50° = 130°.
3. ∠4 and ∠2 are vertical angles, so ∠4 = ∠2 = 150°.
4. ∠5 and ∠6 are supplementary, so ∠5 = 180° − 124° = 56°.
5. ∠4 and ∠8 are corresponding angles, so ∠8 = ∠4 = 140°.
6. ∠2 and ∠6 are corresponding angles, so ∠2 = ∠6 = 112°.
7. ∠4 and ∠2 are vertical angles, so ∠4 = ∠2 = 112°.
   ∠4 and ∠8 are corresponding angles, so ∠8 = ∠4 = 112°.
8. ∠1 and ∠5 are corresponding angles, so ∠1 = ∠5 = 25°.
9. ∠1 and ∠5 are corresponding angles, so ∠1 = ∠5 = 35°.
   ∠5 and ∠8 are supplementary, so ∠8 = 180° − 35° = 145°.
10. ∠4 and ∠8 are corresponding angles, so ∠8 = ∠4 = 146°.
    ∠8 and ∠6 are vertical angles, so ∠8 = ∠6 = 146°.
11. $180° - 27° = 153°$
12. The angle that is vertical with the 62°-angle is also 62°. The other angles are both 180° − 62°, or 118°.

LESSON 22

ON YOUR OWN (page 52): 41°

PAGE 53

1. acute; $180° - (52° + 64°) = 180° - 116° = 64°$
2. acute; $180° - (27° + 73°) = 180° - 100° = 80°$
3. obtuse; $180° - (126° + 25°) = 180° - 151° = 29°$
4. Two of the sides are the same length, 5-in. each. A triangle having two sides of the same length is an isosceles triangle.
5. a scalene triangle
6. $180° - (55° + 90°) = 180° - 145° = 35°$
7. $4 \times 180° = 720°$
8. $35° + 77° = 112°, 180° - 112° = 68°$
9. Answers will vary. Triangle should have three acute angles and no congruent sides.

LESSON 23

ON YOUR OWN (page 54): 135°

PAGE 55

1. ∠A = ∠C, so ∠A + ∠C = 155° + 155° = 310°.
   All the angles in a parallelogram add up to 360°, so this leaves 360° − 310° = 50° for ∠B + ∠D.
   ∠B = ∠D, so ∠B = 25°
2. ∠D = ∠B = 30°
   ∠A = ∠C = $\frac{1}{2}[360° - (2 \times 30°)] = \frac{1}{2}(360° - 60°) = \frac{1}{2} \times 300° = 150°$
3. $\overline{AB} = \overline{CD} = 10$ cm
4. $\overline{DA} = \overline{BC} = 2$ in.
5. 2×9 ft $+ 2 \times 15$ ft $= 18$ ft $+ 30$ ft $= 48$ ft
6. The opposite angle is 45°. The other two angles are equal:
   $\frac{1}{2}[360° - (2 \times 45°)] = \frac{1}{2}(360° - 90°) = \frac{1}{2} \times 270° = 135°$
7. square (rectangle, parallelogram, and quadrilateral could also be correct.)
8. $360° - (25° + 35°) = 360° - 60° = 300°$
   $x + 2x = 3x = 300°$
   $x = 100°; 2x = 200°$

LESSON 24

ON YOUR OWN (page 56): 13 in.

PAGE 57

1. $\overline{CA}$, $\overline{DC}$, or $\overline{CE}$
2. $\overline{OS}$, $\overline{OJ}$, $\overline{OG}$, and $\overline{OT}$ are all radii.
3. $\overline{CD}$, $\overline{CE}$, $\overline{CA}$, and $\overline{AE}$ all have only one endpoint on the edge of the circle.
4. $\overline{AB}$ and $\overline{BE}$
5. 5 chords: $\overline{GS}$, $\overline{GJ}$, $\overline{SJ}$, $\overline{ST}$, $\overline{GT}$
6. $\overline{DE}$ and $\overline{AB}$
7. The angles form a straight line, so they total 180°.
8. $\angle TOS$ and $\angle TOG$ are supplementary, so $\angle TOG = 180° - 86° = 94°$.
9. All radii in a circle are equal, so because $\overline{OG} = \overline{OJ}$, $\angle OJG = \angle OGJ$. The angles on a triangle add to 180°, so $180° - 150° = 30°/2 = 15°$ per angle. $\angle OJG = 15°$
10. $\angle TOS = 86°$, and the other two angles are equal.
    $180° - 86° = 94°$
    $94° \div 2 = 47°$; $\angle OTS = 47°$
11. $d = 2r = 28$ in.
    $r = 14$ in.
12. 180°
13. $10° \times (5.5 \div 0.5) = 10° \times 11 = 110°$
14. $360° \div 5 = 72°$

LESSON 25

ON YOUR OWN (page 58): 25.12 cm

PAGE 59

1. $C = \pi d = 4 \times 3.14 = 12.56$ ft
2. $C = 2\pi r = 2 \times 3.14 \times 5$ mm $= 31.4$ mm
3. $C = \pi d = 3.14 \times 19$ yd $= 59.66$ yd
4. $C = 2\pi r = 2 \times 3.14 \times 2$ mi $= 12.56$ mi
5. $C = \pi d = 3.14 \times 3$ cm $= 9.42$ cm
6. $C = 2\pi r = 2 \times 3.14 \times 6$ in. $= 37.68$ in.
7. $C = 2\pi r = 2 \times 3.14 \times 4.5$ m $= 28.26$ m
8. $d = \frac{C}{\pi} = \frac{18\,\text{ft}}{3.14} = 5.73$ ft
9. $C = \pi d = 3.14 \times 36$ in. $= 113.04$ in.
10. $C = 2\pi r = 2 \times 3.14 \times 5$ cm $= 31.4$ cm
11. $C = \pi d = 3.14 \times 1$ ft $= 3.14$ ft
12. $C = 2\pi r = 2 \times 3.14 \times 16$ ft $= 100.48$ ft

LESSON 26

ON YOUR OWN (page 60): 6 in.^{2}

PAGE 61

1. $A = \frac{1}{2} \times b \times h = \frac{1}{2} \times 4$ ft $\times 5$ ft $= 10$ ft^2
2. $A = \frac{1}{2} \times b \times h = \frac{1}{2} \times 5$ km $\times 2$ km $= 5$ km^2
3. $A = \frac{1}{2} \times b \times h = \frac{1}{2} \times 12$ cm $\times 4$ cm $= 24$ cm^2
4. $h = 2 \times \frac{A}{b} = 2 \times \frac{12\,\text{yd}^2}{8\,\text{yd}} = 3$ yd
5. $h = 2 \times \frac{A}{b} = 2 \times \frac{20\,\text{km}^2}{8\,\text{km}} = 5$ km
6. $b = 2 \times \frac{A}{h} = 2 \times \frac{2.5\,\text{m}^2}{1\,\text{m}} = 5$ m
7. $A = \frac{1}{2} \times b \times h = \frac{1}{2} \times 8$ ft $\times 9$ ft $= 36$ ft^2
8. $A = \frac{1}{2} \times b \times h = \frac{1}{2} \times 8.5$ in. $\times 11$ in. $= 46.75$ in.^{2}
9. $A = \frac{1}{2} \times b \times h = \frac{1}{2} \times 1$ yd $\times 1.5$ yd $= 0.75$ yd^2
10. $A = \frac{1}{2} \times b \times h = \frac{1}{2} \times 8$ m $\times 2$ m $= 8$ m^2
    $A = l \times w = 4$ m $\times 8$ m $= 32$ m^2
    Total area $= 8$ m^2 $+ 32$ m^2 $= 40$ m^2
11. $A = \frac{1}{2} \times b \times h = \frac{1}{2} \times 5$ ft $\times 2$ ft $= 5$ ft^2
    $A = l \times w = 3$ ft $\times 6$ ft $= 18$ ft^2
    Total area $= 18$ ft^2 $+ 5$ ft^2 $= 23$ ft^2
12. $h = 2 \times \frac{A}{b} = 2 \times \frac{60\,\text{ft}^2}{12\,\text{ft}} = 10$ ft

LESSON 27

ON YOUR OWN (page 62): 78.5 m^2

PAGE 63

1. $r = \sqrt{\frac{A}{\pi}} = \sqrt{\frac{12.56\,\text{ft}^2}{3.14}} = \sqrt{4\,\text{ft}^2} = 2$ ft
2. $A = \pi \times r^2 = 3.14 \times (4\,\text{km})^2 = 3.14 \times 16$ km^2 $= 50.24$ km^2
3. $A = \pi \times r^2 = 3.14 \times (20\,\text{yd})^2 = 3.14 \times 400$ yd^2 $= 1{,}256$ yd^2
4. $r = \sqrt{\frac{A}{\pi}} = \sqrt{\frac{36\,\pi\,\text{cm}^2}{\pi}} = \sqrt{36\,\text{cm}^2} = 6$ cm
5. $r = \sqrt{\frac{A}{\pi}} = \sqrt{\frac{25\,\pi\,\text{yd}^2}{\pi}} = \sqrt{25\,\text{yd}^2} = 5$ yd
6. $r = \sqrt{\frac{A}{\pi}} = \sqrt{\frac{49\,\pi\,\text{m}^2}{\pi}} = \sqrt{49\,\text{m}^2} = 7$ m
    $d = 2 \times 7$ m $= 14$ m
7. $A = \pi \times r^2 = 3.14 \times (7\,\text{ft})^2 = 3.14 \times 49$ ft^2 $= 153.86$ ft^2
8. $A = \pi \times r^2 = 3.14 \times (2\,\text{ft})^2 = 3.14 \times 4$ ft^2 $= 12.56$ ft^2
9. $r = \sqrt{\frac{A}{\pi}} = \sqrt{\frac{200\,\text{ft}^2}{3.14}} = \sqrt{64\,\text{ft}^2} = 8$ ft
10. $A = \pi \times r^2 = 3.14 \times (8\,\text{in.})^2 = 3.14 \times 64$ in.^{2} $= 200.96$ in.^{2}

LESSON 28

ON YOUR OWN (page 64): No

PAGE 65

1. *MP* corresponds to *GJ*, and because the figures are congruent, the sides are equal. *MP* = 2 yd

2. *OP* corresponds to *IJ*. Because the figures are congruent, *OP* = 3 yd.

3. *GH* corresponds to *MN*. Because the figures are congruent, *GH* = 5 yd.

4. $\frac{2 \, m}{3 \, m} = \frac{3 \, m}{BC}$; $\overline{BC}$ = 4.5 m

5. $\frac{3 \, m}{4.5 \, m} = \frac{2m}{QR}$; $\overline{QR}$ = 3 m

6. $\frac{3 \, m}{4.8 \, m} = \frac{2m}{MR}$; $\overline{MR}$ = 3.2 m

7. $\frac{1 \, in.}{12 \, in.} = \frac{2}{3}$ in./h

   $\frac{2}{3}$ in. × 12 in. = 1 in. × h

   h = 8 in.

8. 22 in.

9. $\frac{3 \, ft}{15 \, ft}$ = 5 ft/l

   3 ft × l = 15 ft × 5 ft = 75 ft^2

   l = 25 ft

10. Yes, because they are identical in size and shape.

LESSON 29

ON YOUR OWN (page 66): 4

PAGE 67

1. A regular pentagon has a line of symmetry from each vertex (⊥) to the other side. So there are 5 lines of symmetry.

2. 5 lines cross and create 10 new angles.

3. 1

4. None, no lines of symmetry cross.

5. 2

6. depending on the direction of folding, an arrow pointing to the left or the right

7. 1

8. No, it will look the same only for the line of symmetry.

9. Because of the color difference, there is no way you can draw a line of symmetry so that one side will be a mirror-image of the other side.

10. 2

11. 1, 3, 0, 8

12. A, B, C, D, E, H, I, K, M, O, T, U, V, W, X, Y

13. Yes, 2.5 points of the star are on each side.

14. 15

LESSON 30

ON YOUR OWN (page 68): 7 ft

PAGE 69

1. $7^2 = 49$, $24^2 = 576$ and $25^2 = 625$, and $49 + 576 = 625$, so the triangle is a right triangle. Yes.

2. $(12)^2 + (15)^2 = (20)^2$, $144 + 225 \neq 400$; it is not a right triangle.

3. $c^2 = 6^2 + 8^2 = 36 + 64 = 100$, $c = 10$ m

4. $c^2 = 6^2 + 10^2 = 36 + 100 = 136$, $c = 11.66$ mi

5. $a^2 + 48^2 = 50^2$, $a^2 + 2304 = 2500$, $a^2 = 196$, $a = 14$ km

6. $50^2 = 2500$, $100^2 = 10000$, $2500 + 10000 = 12500$
   $\sqrt{12500}$ yd = $50\sqrt{5}$ yd ≈ 110 yd

7. $25^2 + w^2 = 40^2$, $625 + w^2 = 1600$, $w^2 = 975$, $w = 31.2$ in.

8. $c^2 = 90^2 + 90^2 = 8100 + 8100 = 16200$, $c = 127$. From home plate to second base is about 127 ft.

9. $17^2 + 12^2 = 289 + 144 = 433$, $c = 20.8$ ft. No, the 20-foot cord does not reach 20.8 feet.

10. $a^2 + 2^2 = 2.5^2$, $a^2 + 4 = 6.25$, $a^2 = 2.25$, $a = 1.5$ m

11. $a^2 + 30^2 = 40^2$, $a^2 + 900 = 1600$, $a^2 = 700$, $a = 26.5$ ft

LESSON 31

ON YOUR OWN (page 70): M′(−10, 3), N′(−3, 3), O′(−3, −1), P′(−10, −1)

PAGE 71

1. The original coordinates are: $P(2, 4)$, $Q(7, 4)$, $R(5, 7)$; add $(−3, −1)$ to each set of coordinates: $P′(−1, 3)$, $Q′(4, 3)$, $R′(2, 6)$.

2. The original coordinates are: $P(2, 4)$, $Q(7, 4)$, $R(5, 7)$; add $(2, 0)$ to each set of coordinates: $P′(4, 4)$, $Q′(9, 4)$, $R′(7, 7)$.

3. The original coordinates are: $P(2, 4)$, $Q(7, 4)$, $R(5, 7)$; add $(2, 6)$ to each set of coordinates: $P′(4, 10)$, $Q′(9, 10)$, $R′(7, 13)$.

4. The original coordinates are: $P(2, 4)$, $Q(7, 4)$, $R(5, 7)$; add $(0, −3)$ to each set of coordinates: $P′(2, 1)$, $Q′(7, 1)$, $R′(5, 4)$.

5. The translation is $(0 + −14, 0 + 10)$ or $(−14, 10)$

6. $(−3 + −10, 8 + 5)$, $(−13, 13)$; 13 ft left and 13 ft in front of the gate

7. $(30 + 15, 50 + −20)$, $(45, 30)$; 45 ft east, 30 ft north

8. $(2 + 3, 5 + −2)$, $(5H, 3V)$

LESSON 32

ON YOUR OWN (page 72): See student work.

PAGE 73

1. $A'(-1, -3)$
2. $B'(2, -2)$
3. $E'(-4, -3), F'(-4, -4), G'(-1, -4)$
4. $E'(4, 3), F'(4, 4), G'(1, 4)$
5. $A'(-6, -4), B'(-3, -2), C'(-4, -5)$
6. $M'(-4, 5), N'(-4, 3), O'(-6, 3), P'(-6, 5)$
7. You would see your left hand, but it looks like your right hand.
8. Flipping once creates a mirror image. Flipping again creates a mirror image of the mirror image, which is the original image.

LESSON 33

ON YOUR OWN (page 74): See student work.

PAGE 75

1. $A'(5, 11), B'(3, 8), C'(4, 3)$
2. $D'(0, 0), E'(2, 3), F'(-3, 3)$
3. The side resting against the y-axis should rest against the x-axis.
4. The result is a C above the original facing in the opposite direction.
5. 12 hours make up $360°$, 1 hour is $\frac{360°}{12} = 30°$.
   2 hours $= 2 \times 30° = 60°$
6. $180°$ will reverse his or her direction.
7. $\frac{270}{10} = 27, 270 + 27 = 297°$
8. $\frac{180}{12} = 15°$

LESSON 34

ON YOUR OWN (page 76): $A'(8, 10), B'(6, 4), C'(10, 4)$

PAGE 77

1. $C' = (2 \times 2)$ and $(3 \times 2) = (4, 6)$
   $D' = (6 \times 2)$ and $(7 \times 2) = (12, 14)$
   $E' = (6 \times 2)$ and $(3 \times 2) = (12, 6)$
2. $F' = (-6 \times 0.5)$ and $(3 \times 0.5) = (-3, -1.5)$
   $G' = (-3 \times 0.5)$ and $(-3 \times 0.5) = (-1.5, -1.5)$
   $H' = (-6 \times 0.5)$ and $(-6 \times 0.5) = (-3, -3)$
   $J' = (-3 \times 0.5)$ and $(-6 \times 0.5) = (-1.5, -3)$

3. $P' = (2 \times 0.25)$ and $(-3 \times 0.25) = (0.5, -0.75)$
   $Q' = (7 \times 0.25)$ and $(-3 \times 0.25) = (1.75, -0.75)$
   $R' = (3 \times 0.25)$ and $(-6 \times 0.25) = (0.75, -1.5)$
   $S' = (5 \times 0.25)$ and $(-6 \times 0.25) = (1.25, -1.5)$
4. $A' = (-3 \times 0.5)$ and $(4 \times 0.5) = (-1.5, 2)$
   $B' = (-2 \times 0.5)$ and $(2 \times 0.5) = (-1, 1)$
5. $D' = (-4 \times 2)$ and $(5 \times 2) = (-8, 10)$
   $O' = (-3 \times 2)$ and $(2 \times 2) = (-6, 4)$
   $G' = (-6 \times 2)$ and $(4 \times 2) = (-12, 8)$
6. $C' = (4 \times 0.5)$ and $(5 \times 0.5) = (2, 2.5)$
   $A' = (6 \times 0.5)$ and $(5 \times 0.5) = (3, 2.5)$
   $R' = (6 \times 0.5)$ and $(3 \times 0.5) = (3, 1.5)$
   $L' = (4 \times 0.5)$ and $(3 \times 0.5) = (2, 1.5)$
7. $2 \times 8 = 16$ in.
8. $5 \times 6 = 30$ in.

LESSON 35

ON YOUR OWN (page 78): 60 ft^3

PAGE 79

1. $7 \times 3 \times 4 = 84$ m^3
2. $2 \times 3 \times 6 = 36$ mi^3
3. $(\frac{1}{2} \times 2 \times 3) \times 7 = 21$ in.^{3}
4. $2 \times 5 \times 3 = 30$ ft^3
5. $\frac{1}{3} \times (4 \times 4) \times 3 = 16$ in.^{3}
6. $\frac{1}{3} \times (6 \times 5) \times 4 = 40$ cm^3
7. $\frac{1}{3} \times (\frac{1}{2} \times 2 \times 4) \times 3 = 4$ yd^3
8. $(3 \times 3 \times 3) + (\frac{1}{3} \times (3 \times 3) \times 2) = 27 + 6 = 33$ cm^3
9. $\frac{1}{3} \times (3 \times 3) \times 5 = 15$ mm^3
10. $6 \times 4 \times 1.5 = 36$ ft^3
11. $(9 \times 9 \times 2) \times 3 = 486$ in.^{3}
12. $(\frac{1}{2} \times 2 \times 1.5) \times 10 = 15$ in.^{3}
13. $\frac{1}{3} \times (230 \times 230) \times 150 = 2,645,000$ m^3

LESSON 36

ON YOUR OWN (page 80): 75.36 in.^{2}

PAGE 81

1. If the diameter is 2 feet, then the radius is 1 foot. The height is 5 feet, so the surface area is $2 \times (\pi \times (1 \text{ ft})^2) + (2 \times \pi \times 1 \text{ ft}) \times 5 \text{ ft} = 37.68$ ft^2.
2. $2 \times (3.14 \times 4^2) + 2 \times 3.14 \times 4 \times 1 = 100.48 + 25.12 = 125.60$ mi^2
3. $2 \times (3.14 \times (\frac{6}{2})^2) + 2 \times 3.14 \times (\frac{6}{2}) \times 6 = 56.52 + 113.04 = 169.56$ m^2

4. $2 \times (3.14 \times (\frac{10}{2})^2) + 2 \times 3.14 \times (\frac{10}{2}) \times 8 = 157 + 251.2 =$ 408.2 yd^2

5. $2 \times (3.14 \times 4^2) + 2 \times 3.14 \times 4 \times 10 = 100.48 + 251.2 =$ 351.68 in.^{2}

6. $2 \times (3.14 \times (\frac{6}{2})^2) + 2 \times 3.14 \times (\frac{6}{2}) \times 8 = 56.52 + 150.72 =$ 207.24 mm^2

7. $2 \times (3.14 \times (\frac{20}{2})^2) + 2 \times 3.14 \times (\frac{20}{2}) \times (20) = 628 + 1{,}256 =$ 1,884 cm^2 (Remind students to convert 0.2 m to 20 cm.)

8. $2 \times (3.14 \times 1^2) + 2 \times 3.14 \times 1 \times 4 = 6.28 + 25.12 = 31.40$ ft^2

9. $2 \times (3.14 \times 3^2) + 2 \times 3.14 \times 3 \times 20 = 56.52 + 376.8 =$ 433.32 cm^2

10. $r = 1$ in., $h = 6$ in., so SA $= 2 \times \pi \times (1$ in.$)^2 + 2\pi \times (1$ in.$) \times$ 6 in. $= 43.96$ in.^{2}

11. $2 \times (3.14 \times (\frac{18}{2})^2) + 2 \times 3.14 \times (\frac{18}{2}) \times 6 = 162 \times 3.14 + 108$ $\times 3.14 = 847.8$ in.^{2}

12. $(3.14 \times (\frac{10}{2})^2) + 2 \times 3.14 \times (\frac{10}{2}) \times 4 = 78.5 + 125.60$ $= 204.1$ cm^2

13. $(3.14 \times (\frac{2}{2})^2) + 2 \times 3.14 \times (\frac{2}{2}) \times 4 = 3.14 + 8 \times 3.14 =$ 28.26 ft^2

LESSON 37

ON YOUR OWN (page 82): 46 in.^{2}

PAGE 83

1. $2 \times (2 \times 2) + 2 \times (2 \times 2) + 2 \times (2 \times 2) = 8 + 8 + 8 = 24$ m^2

2. $2 \times (5 \times 1) + 2 \times (5 \times 2) + 2 \times (2 \times 1) = 10 + 20 + 4 =$ 34 yd^2

3. $2 \times (1 \times 5) + 2 \times (1 \times 5) + 2 \times (5 \times 5) = 10 + 10 + 50 =$ 70 in.^{2}

4. $2 \times (1 \times 2) + 2 \times (2 \times 3) + 2 \times (1 \times 3) = 4 + 12 + 6 = 22$ ft^2

5. $2 \times (2 \times 3) + 2 \times (3 \times 4) + 2 \times (2 \times 4) = 12 + 24 + 16 =$ 52 mi^2

6. $2 \times (30 \times 10) + 2 \times (30 \times 10) + 2 \times (10 \times 10) = 600 + 600$ $+ 200 = 1{,}400$ m^2

7. 350 in.$^2 = 2 (15$ in. $\times 10$ in.$) + 2 (15$ in. $\times h$ in.$) + 2 (10$ in. $\times$ h in.$)$, so 350 in.$^2 = 300$ in.$^2 + 30$ in. $\times h + 20$ in. $\times h =$ 300 in.$^2 + 50$ in. $\times h$ in. Finally, 50 in.$^2 = 50$ in. $\times h$ in., and so $h = 1$; the height is 1 in.

8. $2 \times (2 \times 2) + 2 \times (2 \times 3) + (3 \times 2) = 8 + 12 + 6 = 26$ ft^2

9. $2 \times (1 \times 10) + 2 \times (10 \times 12) + 2 \times (1 \times 12) = 20 + 240 +$ $24 = 284$ cm^2

10. $2 \times (10 \times 8) + 2 \times (10 \times 8) + (10 \times 10) \times 15 = 160 +$ $160 + 100 - 15 = 405$ ft^2. No, you only have enough paint to cover 400 sq ft, and you need to cover 405 sq ft.

11. $2 \times (10 \times 8) + (10 \times 1) = 160 + 10 = 170$ in.^{2}

12. $2 \times (2 \times 1) + 2 \times (2 \times 1) + 2 \times (1 \times 1) = 4 + 4 + 2 = 10$ ft^2

LESSON 38

ON YOUR OWN (page 84): 50.24 in.^{3}

PAGE 85

1. $(\pi \times 2^2) \times 10 = 125.6$ cm^3

2. $(\pi \times 2^2) \times 6 = 75.36$ ft^3

3. $(\pi \times 3^2) \times 5 = 141.3$ yd^3

4. $(\pi \times (\frac{12}{2})^2) \times 10 = 1{,}130.4$ km^3

5. $C = 10 \times \pi = 2 \times \pi \times r, r = 5; V = (\pi \times 5^2) \times 5 = 392.5$ cm^3

6. $C = 6 \times \pi = 2 \times \pi \times r, r = 3; V = (\pi \times 3^2) \times 15 = 423.9$ mm^3

7. $(3.14 \times 3^2) \times 10 = 282.6$ cm^3

8. $(3.14 \times 6^2) \times 10 = 1{,}130.4$ in^3

9. $(3.14 \times 4^2) \times 1000 = 50{,}240$ m^3

10. $(3.14 \times (\frac{3}{2})^2) \times 10 = 70.65$ in.^{3}

11. $(3.14 \times (\frac{12}{2})^2) \times 15 = 1{,}695.6$ in.^{3}

12. $(3.14 \times (\frac{6}{2})^2) \times 14 = 395.64$ cm^3

LESSON 39

ON YOUR OWN (page 86): 47.1 cm^3

PAGE 87

1. $\frac{1}{3} \times (3.14 \times 1^2) \times 3 = 3.14$ km^3

2. $\frac{1}{3} \times (3.14 \times 2^2) \times 6 = 25.12$ m^3

3. $\frac{1}{3} \times (3.14 \times 3^2) \times 4 = 37.68$ in.^{3}

4. $\frac{1}{3} \times (3.14 \times (\frac{8}{2})^2) \times 15 = 251.2$ ft^3

5. $\frac{1}{3} \times (3.14 \times (\frac{4}{2})^2) \times 9 = 37.68$ mi^3

6. $C = 6 \times \pi = 2 \times \pi \times r, r = 3; V = \frac{1}{3} \times (3.14 \times 3^2) \times 10 =$ 94.2 mm^3

7. $\frac{1}{3} \times (3.14 \times 2^2) \times 12 = 50.24$ cm^3

8. $\frac{1}{3} \times (3.14 \times 12^2) \times 20 = 3{,}014.4$ mm^3

9. $\frac{1}{3} \times (3.14 \times (\frac{1}{2})^2) \times 3 = 0.785$ ft^3

10. $\frac{1}{3} \times (3.14 \times (\frac{4}{2})^2) \times 3 = 12.56$ in.^{3}

11. $C = 4 \times \pi = 2 \times \pi \times r, r = 2; V = \frac{1}{3} \times (3.14 \times 2^2) \times 24 =$ 100.48 in.^{3}

12. $\frac{1}{3} \times (3.14 \times (\frac{6}{2})^2) \times 5 = 47.1$ ft^3

LESSON 40

ON YOUR OWN (page 88): 6 in.

PAGE 89

1. $V = \frac{1}{3}Bh$
$B = 3 \times 4 = 12$
$V = \frac{1}{3}(12 \times 6)$
$V = 24$ in.^{3}

2. $V = \frac{1}{3}Bh$
$B = 7.5 \times 5 = 37.5$
$V = \frac{1}{3}(37.5 \times 10)$
$V = 125$ cm^3

3. $V = \frac{1}{3}Bh$
$B = \frac{1}{2} \times 2 = 1$
$V = \frac{1}{3}(1 \times 3)$
$V = 1$ ft^3

4. $V = \frac{1}{3}Bh$
$B = 20 \times 7 = 140$
$V = \frac{1}{3}(140 \times 12)$
$V = 560$ in.^{3}

5. $V = \frac{1}{3}\pi r^2 h$
$V = \frac{1}{3} \times (3.14 \times 8^2) \times 16$
$V = \frac{1}{3} \times (3.14 \times 64) \times 16$
$V = \frac{1}{3} \times (200.96 \times 16)$
$V = \frac{1}{3} \times 3,215.36$
$V = 1,071.79$ cm^3

6. $V = \frac{1}{3}\pi r^2 h$
$V = \frac{1}{3} \times (3.14 \times 2^2) \times 3$
$V = \frac{1}{3} \times (3.14 \times 4) \times 3$
$V = \frac{1}{3} \times (12.56 \times 3)$
$V = \frac{1}{3} \times 37.68$
$V = 12.56$ m^3

7. $V = \frac{1}{3}\pi r^2 h$
$V = \frac{1}{3} \times (3.14 \times 0.5^2) \times 4$
$V = \frac{1}{3} \times (3.14 \times 0.25) \times 4$
$V = \frac{1}{3} \times (0.785 \times 4)$
$V = \frac{1}{3} \times 3.14$
$V = 1.05$ ft^3

8. $V = \frac{1}{3}\pi r^2 h$
$V = \frac{1}{3} \times (3.14 \times 1^2) \times 6$
$V = \frac{1}{3} \times (3.14 \times 1) \times 6$
$V = \frac{1}{3} \times (3.14 \times 6)$
$V = \frac{1}{3} \times 18.84$
$V = 6.28$ yd^3

9. $V = \pi r^2 h$
300 in.$^3 = 3.14 \times 3^2 \times h$
300 in.$^3 = (3.14 \times 9) \times h$
300 in.$^3 = 28.26 \times h$
$h = 300 \div 28.26$
$h = 10.62$ in.

10. $V = l \times w \times h$
3 in$^3 = 3 \times 2 \times h$
$6h = 3$
$h = 3 \div 6$
$h = 0.5$ in.

11. $V = l \times w \times h$
$V = 5 \times 4 \times 3$
$V = 60$ ft^3
60 ft$^3 > 50$ ft^3
so the tank is large enough.

12. $V = \frac{1}{3}\pi r^2 h$
$V = \frac{1}{3} \times 3.14 \times 1.5^2 \times 6$
$V = \frac{1}{3} \times 3.14 \times 2.25 \times 6$
$V = \frac{1}{3} \times 42.39$
$V = 14.13$ in.^{3}

LESSON 41

ON YOUR OWN (page 90): c

PAGE 91

1. A shows the view of the object from the left side.
2. C shows the view of the object from the front.
3. B shows the view of the object from above.
4. D shows the view of the object from the right side.
5. C shows the view of the object from the left side.
6. B shows the view of the object from the front.
7. A shows the view of the object from above.
8. B shows the view of the object from the left side.
9. B shows the view of the object from the front.
10. A shows the view of the object from above.

Most Often Used Measurement Units

English System

Length

1 foot (ft or ')	= 12 inches (in. or ")
1 yard (yd)	= 3 feet
	= 36 inches
1 mile (mi)	= 1,760 yards
	= 5,280 feet

Weight

1 pound (lb)	= 16 ounces (oz)
1 ton (tn)	= 2,000 lb

Capacity

1 tablespoon (tbsp)	= 3 teaspoons (tsp)
1 cup (c)	= 8 fluid ounces (fl oz)
1 pint (pt)	= 2 cups
	= 16 fluid ounces
1 quart (qt)	= 2 pints
	= 4 cups
	= 32 fluid ounces
1 gallon (gal)	= 4 quarts
	= 128 fluid ounces

Metric System

Length

1 centimeter (cm)	= 10 millimeters (mm)
1 meter (m)	= 100 cm
	= 1,000 mm
1 kilometer (km)	= 1,000 m

Weight

1 gram(g)	= 1,000 milligrams (mg)
1 metric ton (t)	= 1,000 kg
1 kilogram (kg)	= 1,000 g

Capacity

1 metric tablespoon	= 3 metric teaspoons
	= 15 milliliters (mL)
1 metric cup	= 250 milliliters
1 liter	= 1,000 milliliters
	= 4 metric cups
1 kiloliter (kL)	= 1,000 liters

Converting Between Customary Units and Metric Units

Converting from Customary Units to Metric Units

Length

1 inch	=	2.54 centimeters
1 foot	=	0.305 meters
1 yard	≈	0.91 meters
1 mile	≈	1.6 kilometers

Weight

1 ounce	≈	28 grams
1 pound	≈	0.45 kilogram
1 ton	≈	0.91 metric ton

Capacity

1 teaspoon	≈	4.9 milliliters
1 cup	≈	0.94 metric cup
1 fluid ounce	≈	30 milliliters
1 quart	≈	0.94 liter
1 gallon	≈	3.8 liters

Converting from Metric Units to Customary Units

Length

1 centimeter	=	0.394 inches
1 meter	=	3.28 feet
1 meter	≈	1.09 yards
1 kilometer	≈	0.6 miles

Weight

1 gram	≈	0.035 ounces
1 kilogram	≈	2.2 pounds
1 metric ton	≈	1.10 tons

Capacity

1 milliliter	≈	0.204 teaspoon
1 metric cup	≈	1.06 cups
1 milliliter	≈	0.033 fluid ounces
1 liter	≈	1.06 quarts
1 liter	≈	0.26 gallons

Important Formulas and Common Geometric Shapes

Perimeter	
Square	$P = 4s$
Rectangle	$P = 2l + 2w$
Circumference of a circle	$C = \pi d$ or $C = 2\pi r$
Area	
Square	$A = s^2$
Rectangle	$A = lw$
Parallelogram	$A = bh$
Triangle	$A = \frac{1}{2}bh$
Circle	$A = \pi r^2$
Volume	
Cube	$V = s^3$
Rectangular solid	$V = lwh$
Cylinder	$V = \pi r^2 h$
Pythagorean theorem	$a^2 + b^2 = c^2$

Square

$P = 4s$

$A = s^2$

Triangle

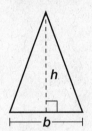

$A = \frac{1}{2}bh$

Circle

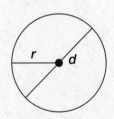

$A = \pi r^2$

$C = \pi d$ or $C = 2\pi r$

Rectangle

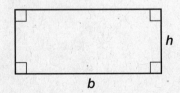

$P = 2b + 2h$

$A = bh$

Trapezoid

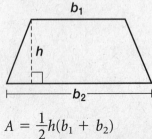

$A = \frac{1}{2}h(b_1 + b_2)$

Parallelogram

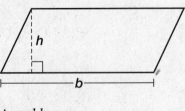

$A = bh$

Three-Dimensional Geometric Figures and Formulas

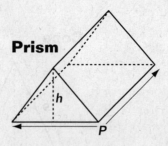

Prism

$LA = ph$

$SA = LA + 2B$

$V = Bh$

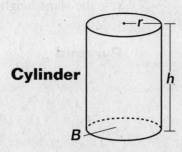

Cylinder

$LA = 2\pi rh$

$SA = LA + 2B$

$V = Bh$ or $V = \pi r^2 h$

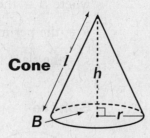

Cone

$LA = \pi rl$

$SA = LA + B$

$V = \frac{1}{3}bh$ or $V = \frac{1}{3}\pi r^2 h$

LA = Lateral area

SA = Surface area

V = Volume

Blackline Master
Measurement and Geometry, SV 0437-9

Formulas for Surface Area

The surface area of a space figure is the sum of all the faces of the figure. Following are the formulas for finding the surface area of several figures.

Cube

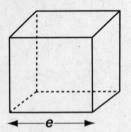

$$SA = 6e^2$$

where e = the length of an edge.

Prism

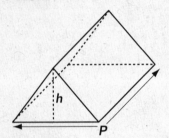

$$SA = 2B + Ph$$

where B = the area of the base,

P = the perimeter of the base, and

h = the height of the prism.

Cylinder

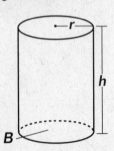

$$SA = 2\pi r(r + h)$$

where r = the radius of the circle and

h = the height of the cylinder.

Cone

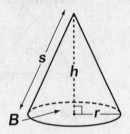

$$SA = \pi r^2 + \pi rs$$

where r = the radius of the circle and

s = the slant height.

Pyramid

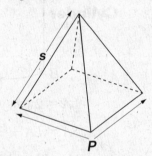

$$SA = B + \frac{1}{2}Ps$$

where B = the area of the base,

P = the perimeter of the base, and

s = the slant height of the lateral faces.

Sphere

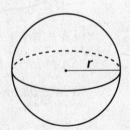

$$SA = 4\pi r^2$$

where r = the radius of the sphere.

Pythagorean Theorem

A special relationship exists between the sides in a right triangle. This relationship is named for the Greek mathematician Pythagoras. The special formula, known as the **Pythagorean theorem,** helps you find distances between points on a coordinate plane, find missing triangle side lengths, or determine if a triangle has a right angle.

In a right triangle, the sides of the triangle that form the right angle are called the *legs* of the triangle. The side opposite the right angle is called the *hypotenuse*.

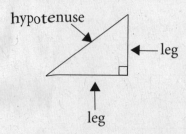

The formula for the Pythagorean theorem is: $a^2 + b^2 = c^2$

If three numbers work in the formula, then you know you have a right triangle.

Coordinate Plane

You can use this blank coordinate plane to help you plot points and graph functions.

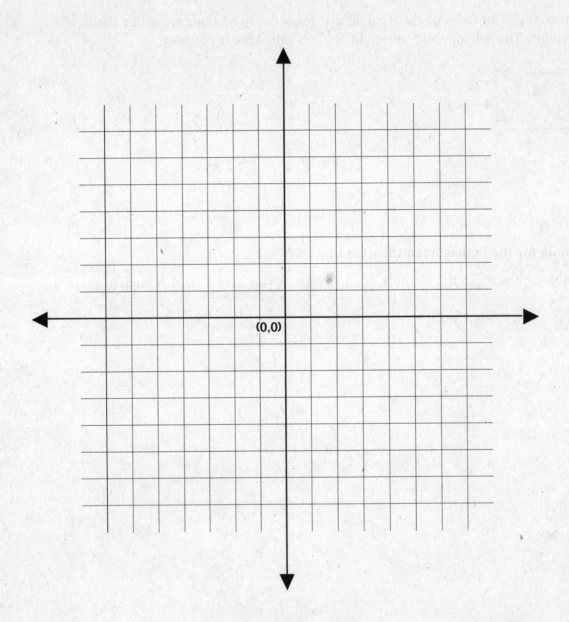

Blackline Master
Measurement and Geometry, SV 0437-9